Uncertainty-Aware Complex System State Reconstruction for Resilient Modern Power Systems

Yoru Takeshi

Uncertainty-Aware Complex System State Reconstruction for Resilient Modern Power Systems

First Edition February 2025

Copyright © 2025 Yoru Takeshi

Written by Yoru Takeshi

CONTENTS

CHAPTER 1 INTRODUCTION	**1-32**
1.1 Introduction	1
1.2 Technologies for Smart Grid applications	6
1.2.1 Advanced Metering Infrastructure	6
1.2.2 Phasor Measurement Units and Wide Area Protection and Control	10
1.2.3 Intelligent Electronic Devices	17
1.2.4 Two way Communication Technologies	20
1.3 Wide Area Monitoring	24
1.4 Smart Grid Distribution	29
1.5 Conclusion	31
CHAPTER 2 LITERATURE SYRVEY	**33-71**
2.1 Introduction	33
2.1.1 Demand Based Response	35
2.1.2 Incentives for Demand Response	35
2.1.3 Strategies for Demand Response	37

2.1.4 Domestic Load Classification	37
2.2 Smart Grid	39
2.3 Parameter Estimation	46
2.4 Phasor Measurement Unit	47
2.4.1 Power System Monitoring with PMU	48
2.4.2 Analytics and Control of PMU	49
2.4.3 Optimized PMU placement in PMU	50
2.4.4 PMU for State Estimation	50
2.4.5 PMU Application for Assessment of Voltage Stability	51
2.4.6 Fault Detection/Location with PMU Signal	51
2.4.7 PMU Application for Control and Monitoring	52
2.5 Kalman Filter	59
2.6 Optimization Techniques	61
2.7 Inference out of the Literature Review	67
2.8 Problem Formulation	68
2.9 Research Objectives	69
2.10 Outline of thesis	69
2.11 Conclusion	70
CHAPTER 3: KALMAN FILTER BASED PARAMETER ESTIMATION	**72-97**
3.1 Introduction	72
3.1.1 Kalman Filter	75
3.1.2 Ensemble Kalman Filter	77
3.1.3 Adaptive Kalman Filter	77
3.1.4 Extended Kalman Filter	78
3.1.5 Unscented Kalman Filter	81

3.2 Observational Comparative Performance	84
3.3 IEEE 30 Bus System	86
3.4 Performance of Kalman Filter	87
3.5 Performance of Extended Kalman Filter	89
3.6 Performance of Unscented Kalman Filter	92
3.7 Experimental Comparative Performance	95
3.8 Conclusion	97
CHAPTER 4: KALMAN FILTER ESTIMATION WITH BAYESIAN APPROACH	**98-110**
4.1 Kalman Procedure	98
4.2 IEEE 30 Bus System	99
4.3 Performance of Kalman Filter using Taylor Expansion	99
4.4 Performance of Kalman Filter using Bayesian Approach	101
4.5 Performance of Kalman Filter using Bayesian with Taylor expansion	106
4.6 Comparative Performance of Kalman-Taylor, Kalman-Bayesian and Kalman-Taylor-Bayesian	107
4.7 Conclusion	110
CHAPTER 5: PARTICLE SWARM OPTIMIZATION	**111-125**
5.1 Optimization Techniques	111
5.2 Particle Swarm Optimization	111
5.3 Kalman Bayesian Taylor PSO	114
5.4 IEEE 30 Bus System	116

5.5 Performance of Kalman PSO	116
5.6 Performance of Kalman Taylor PSO	118
5.7 Performance of Kalman-Bayesian-Taylor PSO	120
5.8 Comparison of Hybrid Approach es	122
5.9 Conclusion	125
CHAPTER 6: EXTENDED KALMAN FILTER	**126-140**
6.1 Introduction	126
6.2 IEEE 30 Bus System	126
6.3 Performance of Extended Kalman Filter	126
6.4 Performance of Extended Kalman Filter with Bayesian Approach	129
6.5 Performance of Extended Kalman Filter with Particle Swarm Optimization	130
6.6 Comparative Performance	
6.6.1 Comparative Performance of EKF, EKF-Bayesian and EKF PSO in small time frame	133
6.6.2 Comparative Performance of EKF, EKF-Bayesian and EKF PSO in large time frame	134
6.6.3 ComparativePerformance of EKF, Bayesian -Kalman and PSO-Kalman	136
6.6.4 Comparative Performance of EKF, EKF-Bayesian and KF-Bayesian	137
6.6.5 Comparative Performance of Kalman filter using different approach	137
6.7 Conclusion	139
CHAPTER 7: CONCLUSION AND FUTURE SCOPE	**141-145**
7.1 Conclusion	141
7.2 Future Scope	145

| References | 146-157 |

Chapter1 INTRODUCTION

1.1 Introduction

The smart grid at its core represents the use of rising technology in order to support the energy and the cost-based efficiency. A smartly designed energy network, reads in an automatic way and reacts to the changes of supply as well as the demand. It offers a large potential for maintenance of security of the supply system efficiently. When these are linked or coupled with the smart meter roll-out, then the possible efficiency is always larger as the customers easily adapt with their own demands on real time basis and usually increase the renewable energy integration into the grid [1]. Keeping it in mind, a target has been set by the European Union(EU) to change around 80 percent of the already existing meters of electricity by 2020 so as to bring about a possible reduction of emission across the EU to about 9 percent and the same reduction in case of annual consumption of ordinary energy. The ambitions of EU were basically set out in innovation-led electricity-based system transformation and technology-based context. According to the reports of World Economic Forum three significant trends have been identified that are going to rattle large conventional structures from the generation process beyond the smart meter [2] . These are:

- Economical large sectors electrification such as heating and transport;

- The process of decentralisation, propelled by an acute fall in distributed form of energy resources cost like distributed generation, distributed storage, energy efficiency and flexible demand;

- The process of digitalisation of grids, with smart sensors, smart metering, process of automation and other technologies based on digital type of network like arrival of Internet of Things (IoTs) and a large power surge consuming associated devices.

Demonstrating [3] how the components connected complement other type of elements in the process of development, the World Economic Forum (WEF) expresses that the process of electrification is complex for long term goals in reducing the carbon and it will present relevant forms of renewable energy sharing process. The process of decentralisation makes a significant coordination and forms the active elements of the system with respect to the consumers [4]. Digitalisation helps in supporting both the trends by sanctioning the addition

INTRODUCTION

of control, which includes a real-time, automatic optimized production and consumption and the process of interaction with consumers of the system. These modernized trends are usually run by a large number of factors, whereas, the most noticeable models and, the statement provided by the WEF states, "the considerable advancement to the electrical system-based utilisation of asset rates", using an example of electric cars of a region for a greater potential development and growth [5] [6]. The economic and societal integration benefits of such trends also range on wide basis. The prior WEF analysis provides estimation of data that the process of transformation of electricity over the next decade might help in generation of $2.4 billion value. There are economic benefits like creation of job and societal advantages such as the improvement in environmental conditions.

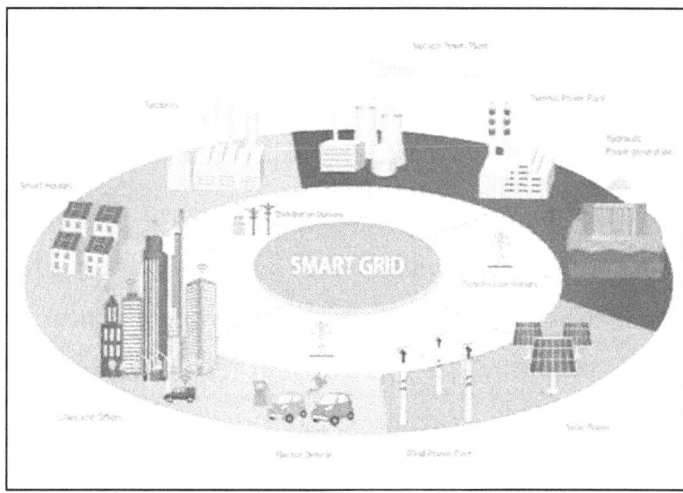

Figure 1.1 Smart Grid [3][70]

The Smart Grid for the most part conforms to an enhanced electricity supply chain i.e. driven from the significant power plant resulting in a proficient route into our houses. In US, the number of power plants is huge which are helping in the generation of electricity like coal, nuclear, wind, hydro and countless other sort of resources. The station that plays the role of generation helps in producing the electric energy at a predefined voltage of electricity. Such type of voltage is stepped up or further increased to high voltages that is around 500,000 volts, so as to increase power transmission efficiency over enormous areas based on distance. When the power is accessible close to the town, at that point the electrical voltage is stepped-

INTRODUCTION

down to a lower distribution voltage. When this electrical power arrives near the consumers premises, it is again stepped-down to voltage levels to be utilized in the household framework with the help of another transformer [7] [22]. This power arriving at the house is metered through the users electrical energy meter. The voltage of the clients premises is around 220-240 volts for most of the machines used at home. In most regions, the electricity-based system clarified above is worn out. Moreover, the development of the populace in the vast majority of the regions has commonly made the whole transmission system to be overused and delicate. There is an increase in the electronic gadgets in household frameworks, like microwave ovens, highly-defined television computers, wireless telephones, and even electronic controls with respect to dishwashers, fridges, and ovens. These recent gadgets are more vulnerable to the variations to the voltage than more established incandescent bulbs, and engines (motors). Unfortunately, today's electrical grid framework is getting progressively frail or weak. The household appliances are getting increasingly sensitive to the rapid changes in the electrical framework. The reliability of electrical power will further diminish if some steps are not taken immediately. The expansion of recently built transmission lines will assist the utilities for getting electrical power from the power plants or the stations to the client home. But mostly a large portion of communities do not need new power lines in their provincial territories [8] [28]. Also, the including of new limit or the capacity, however required, won't increase the reliability of old electrical equipment. What is essential now is an innovative methodology that normally helps in expanding the whole productivity of electrical delivery framework. Such a technology will enhance the quality of the system and would also protect the environment by decreasing the emissions of green house gases. This is the essence of Smart Grid technology.

The major concept behind a smart grid is the addition of analysis, monitoring, control, and capabilities of communication to the electrical system for delivery in order to maximize the systems' throughput while reducing the consumption of the energy [9]. The technology of Smart Grid will help in allowing the utilities to drive the system-based electricity efficiently around them and makes the system economically possible. It will further allow the business and homeowners to use the systems' electricity economically.

INTRODUCTION

Table 1.1 Conventional vs. Smart Grid

	CONVENTIONAL GRID SYSTEM	**SMART GRID SYSTEM**
Metering Operation	Solid State, Electromechanical	Microprocessor/Digital
Communication	Either One way or Two way communication	Integrated/Global two-way communication
Interaction of customer	Limited	Extensive
Control system	Limited contingencies of control systems	Pervasive control system
System Reliability	Cascading outages, Estimated: prone to failures	Predictive: Pro-active real time islanding and protection
Topology	Radial	Network
Generation	Centralized	Distributed and Centralized generation
Monitoring	Blind	Self-Monitoring
Maintenance & operation	Check equipment manually	Monitor equipment remotely
Restoration	Manual	Self-healing, Automated
Control of power flow	Control systems, Limited protection, and monitoring	Adaptive protection, WAMPAC.

In the summer time, one may try to keep the home set at around 75 degrees F if the prices are less, but one may be willing to raise the thermostat to about 78 degrees F when the prices are quite high. So, one has the capability to make a flexible choice to manage the electricity use so as to minimize the costs. The technology of Smart Grid is generally built around a large number of technologies that are already in use by the electrical utilities but it further adds the control and communication capabilities that will help in optimizing the entire smart grid operation [31]. Smart Grid is usually positioned and placed to take the benefit of new kind of technologies like electric plug-in hybrid vehicles, distribution automation, several distributed generation forms, smart metering, solar-based energy, system of managing lighting, and many other kinds of technologies.

Why are smart grids necessary?

The conventional system of the grids generally operates in a same corresponding way as 100 years ago, which means that the flow of energy over grid system starting from the central

INTRODUCTION

power plants to the end users, and the system reliability is assured by the process of preservation of surplus amount of capacity. The analysis of such kind of philosophy represents an extravagant and incompetent system which presents the primary form of greenhouse gas emitter and customer of the fossil fuels [9] [24]. Nonetheless, the use of fossil fuels derives to form an energy source of dominant nature, preferably in the countries which are largely industrialised. The already existing topologies of the grid are not at all well suited to distributed, wind energy sources and solar renewable sources, as the arbitrary and generally intermittent kind of nature of such sources usually form various issues in the grid system that does not support the disseminated information to rapidly control the centres [10].

Recently, there has been a lot of paradigm on how the energy electricity gets generated, transmitted, and consumed by the users. The aging grid system is also facing new exceptional challenges usually caused by increased non-linear and digital loads and higher demands of such loads. The digital equipments found in customer electronics and the data centres intermittent response to the outages has redefined the reliability.

The continued improvement and economic development in human life quality is mostly dependent over the access to reliable and affordable electricity. However, [10] [13] most of the already existing grids face a large number of infrastructural issues on the basis of the fact that the energy systems are normally unfit and outdated to face the challenge of rising demand. So, as a result, the process of network congestion occurs more on frequent basis as the capability to quickly react does not take place or exist. Conclusively, these drawbacks may result in subsequent domino effect and blackouts due to communicational lag in between the control centres and the grid system.

Presently, the customers do not have access to the information based on real-time system that might empower them to take optimized decisions in reference to the power usage corresponding to the market-based scenario. It could empower the customers to lower the process of energy consumption during the sharp peak hours that are considered to be most expensive and they could take benefit of reducing the tariffs during the interval of off-peak.

INTRODUCTION

Smart Grid usually represents a possible solution addressing all the above mentioned activities. Truly speaking, a completely implemented task of revamping the grid will help to utilize the renewable energy sources well too. More significantly, the smart grid technology will allow customers to control their energy bills on the basis of real time and will facilitate the charging process of large-scale electric vehicles too

Smart grid challenges

There are significant challenges for accomplishing a transition to another new electrical power system[11]. Initially, electricity was perceived to a great extent as a commodity which made it difficult to engage the client. Furthermore, the framework of regulatory policies also provide a hurdle for distributed resources. Thirdly, the uncertainty around a portion of the principles helps in keeping away noteworthy partners (stakeholders) from making any significant contribution to empower the infrastructure [11] [81]. Lastly an apprehension for newly built business models and a resistance to change makes the implementation of a Smart Grid difficult.

1.2 Technologies for Smart Grid Application

The Smart Grid transition processes are different from the previously existing grids; it needs to be revamped using new technologies of sensing, control, estimation, automation advancements. As a rule, the innovations are expected to appropriately realize the huge fundamental requirements of Smart Grid framework that consists of the following:

1.2.1 Advanced Metering Infrastructure (AMI): This is an important area of Smart Grid technological innovation which is currently in operation in different network systems. For the most part, the AMI can be portrayed as a two-way communicational system and presents the integration of smart sensors, meters, monitoring or observing frameworks, and data management systems which makes possible the assortment and scattering of data between the customers and their utility meters.

The technology of AMI generally represents a unified term for the purpose of description for the global infrastructure from the Energy Smart Meter to a two way network communication for controlling the system-based equipment and all the related system applications on global

INTRODUCTION

basis, which permit the transferring process as well as the usage of energy information gathering on real-time basis [13]. The AMI technology generally creates a two-way communication with probable consumers and thereby presenting the backbone of the grid technology. AMI goal is to get the remote meter readings for the inexact data type, problems related to load-based profiling, network-based identification, partial load curtailment of energy and load audit as an alternative of load shedding/dropping operation.

AMI based building blocks

AMI consists of several components of hardware and software that play an important role in measurement of transmitting information and energy consumption about water, gas, and energy usage to the utility-based customers and companies [14] [30]. The overall technological AMI components include:

- Smart Meters: The advanced devices of smart meter which have the capability in terms of system capacity to gather data about energy, water, and usage of gas at several different intervals of time and transmits the data with the help of constant network of communication to the system-based utility along with the received informational data like utility-based signal pricing and transmitting it to end user.

- Network Communication: The progressive communicational network that helps to support the process of two-way communication helps in enabling the informational data from the smart grid meters to utility-based firms [15] [61]. Various kind of networks like the Power line-based communications, Broad-band over Power-Line i.e. BPL, public networks or Fixed Radio Frequency for example cellular, landline, paging, etc. and fibre optic communication are mainly used for such kind of purposes.

- Meter-based Acquisition of data : The software-based applications on control centre hardware and the DCUS i.e. Data Concentrator Units that are used for bringing in informational data from the smart-meters via the network of communication and transmits it further to the MDMs i.e. meter data management system.

- Meter Data Management System: MDMS Hosting system that stores, receives, and analyses the information of metering.

INTRODUCTION

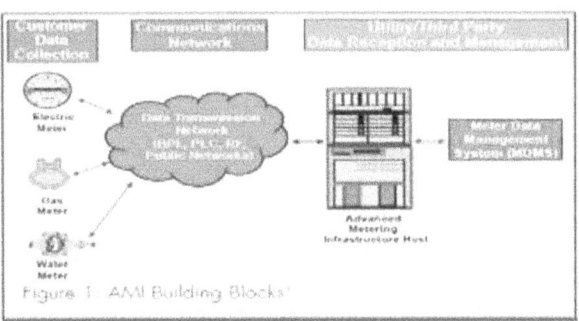

Figure 1.2 AMI components [104]

Home Area Network: Popularly known as HAN, which represents the AMI extension that is usually placed at customer boundary in order to provide facilitation of home appliances structure of communication with the technology of using AMI and hence, it enables a good control of load by both the customer and utility.

The advantages/benefits of utilizing the AMI technology can be commonly classified as the following:

- Operational Benefits : The AMI technology helps the entire grid framework as it makes the data more accurate for meter reading, detection of power theft and blackout reaction, and at the same time eliminates the prerequisites for reading meter on site.

- Financial Benefits : AMI gives monetary advantages to the gas, water and utility firms by bringing down the maintenance costs and also the hardware required to empower the quick restoration during situations of line-blackouts and assists the entire billing process..

- Advantages for customer: This provides several advantages to the consumers as it is able to detect meter failures quickly, restores services fast, and provides accuracy in billing. Finally, this process helps consumer save money while managing energy consumption. At long last, the procedure of AMI encourages the customers to save and manage the money alongside the management of consuming energy.

INTRODUCTION

- <u>Benefits of Security:</u> The AMI also helps in strengthening the framework by reducing the potential dangers to the grid framework due to terrorist attacks and cyber crimes

<u>Challenges</u>

Notwithstanding its far-reaching advantage, the development process of AMI suffers on account of three major challenges comprising of high up-front investment cost, integration process with additional type of grid networks, and the procedure of standardization.

- <u>High Capital Costs:</u> The full-scale deployment based on AMI usually needs to incur an expenditure for each segment of hardware and software frameworks, that incorporate meters, network management software and infrastructure, alongside cost related with maintenance and installation of meters . [17] [26].

- <u>Integration:</u> AMI presents a typically built technological framework that must be incorporated with utilities' IT frameworks, together with Distribution Automation System (DAS), Outage Management Systems (OMS), Work Management System (WMS), Customer Information Systems (CIS), SCADA/DMS, and so on.

- <u>Standardization:</u> The benchmarks of interoperability should be characterized which encourages a consistent requirement for AMI innovation , its processing in general and its deployment activities. This is a key for effective maintenance and connection of an AMI grid connected system.

<u>AMI based Indian Context</u>

Modernized Indian grid may also gain from AMI technology in order to overcome a large number of strains due to rising demand of resources such as gas, electric and water. Specifically, the concept of AMI will help in improving three of the significant features of Indian grids:

INTRODUCTION

- <u>System-based Reliability:</u> The distributors of electricity distinguish and naturally react to the demand of electricity by utilizing AMI innovation. The technology of AMI additionally advances the overall reliability and distribution of the electric power, which brings about limiting power blackouts.
- <u>Energy Costs:</u> Enhanced reliability and diminished power outages with well organized billing tasks will drastically reduce expenses or costs related with maintaining and providing the grid network, in this manner fundamentally bringing down power rates.
- <u>Theft of Electricity:</u> The concept of electricity theft represents a common issue in Indian grid system. The AMI based systems tracking the power usage will help to monitor energy on real time basis thus leading to rising transparency of the system.

1.2.2 Phasor Measurement Units (PMU) and Wide Area Monitoring Protection and Control (WAMPAC)

A phasor diagram represents sinusoidal wave mathematically. Time is taken as reference to determine the phase angle at a known frequency. The values of phasor that represent sinusoidal waveforms of power system in reference to the given universal time coordinates (UTC) and power system frequency are known as Synchro phasors. System frequency, instant of measurement, and waveform determines the phase angle of a synchro phasor. Thus, the phase angles of power system can be measured accurately throughout a PS with a precise universal reference of time. This can also be achieved economically by global positioning system (GPS) technology. By using GPS technology, the main advantage is that synchronization can be accurately detected by its receiver. Synchronized phasor measurements are provided by a device known as PMU. The broadly distributed PMUs in PS framework might be utilized for [9] [10] [25]: controlling the generation of distributed nature, voltages and angular stability, congestion management, Post-Mortem analysis based on disturbances and faults, state estimation protection and control, real time monitoring. PMUs use digital signal processing concepts for estimation of AC waveforms and subsequently changing them into the relevant phasors, in line with the system-based frequency and synchronization of such estimations by the GPS reference-based controlling source. The phasor algorithm is then utilized for processing and sampling of these analog signals for the generation of voltage and current phasors [16]. The basic components of a PMU unit are represented in the figure 1.3.

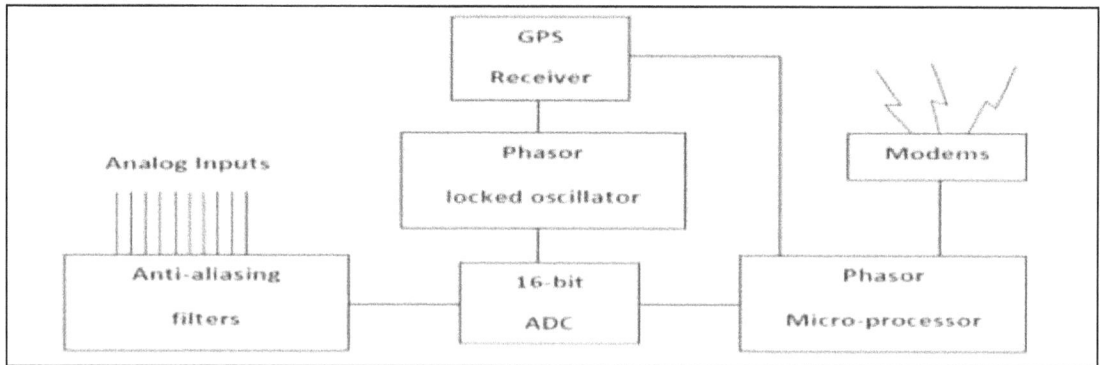

Figure 1.3 PMU (Different Components) [66]

The initial commercial form of PMU i.e. the Macrodyne 1690 generally performed the function of data recording only and was introduced in 1991. PMUs that had the ability of real time measurement were developed by the year 1997.The PMUs which are presently used help in providing a data rate of samples 6-60 per second. In this measurement range, the higher data range may cover the local type of oscillations, actions of generator shafts etc. while the range on the lower end may present the dynamics of power system inter area region.

The data samples are used for estimation of parameters of the phasor with the help of algorithms in order to compute phasors from measured signals. Only the angle of the phasor and magnitude are computed by simple algorithm which assumes a fixed nominal frequency. More accurate results are provided by the more upgraded form of algorithms which help in estimating all the three types of parameters. The most widely used technique of phasor estimation is Discrete Fourier Transform. The other type of methods like the artificial neural networks or Kalman filter help in estimating the phasor diagram. At the control centre is placed a Phasor Data Concentrator (PDC) which is provided with information from other PMUs placed in premises of the smart grid. The PDC will wait for the slowest range data to arrive which it assembles and further aides in the process of sorting the data by a time stamp

INTRODUCTION

technique till the time the slowest range of data reaches. At the control centre, the concentrated data or information by the PDC is additionally used for different applications. Presently, the super PDC collects the information or data from a few PDCs that are usually distributed throughout a specific region. The theoretical measurement diagram for WAM which is PMU based is depicted in figure.1.4

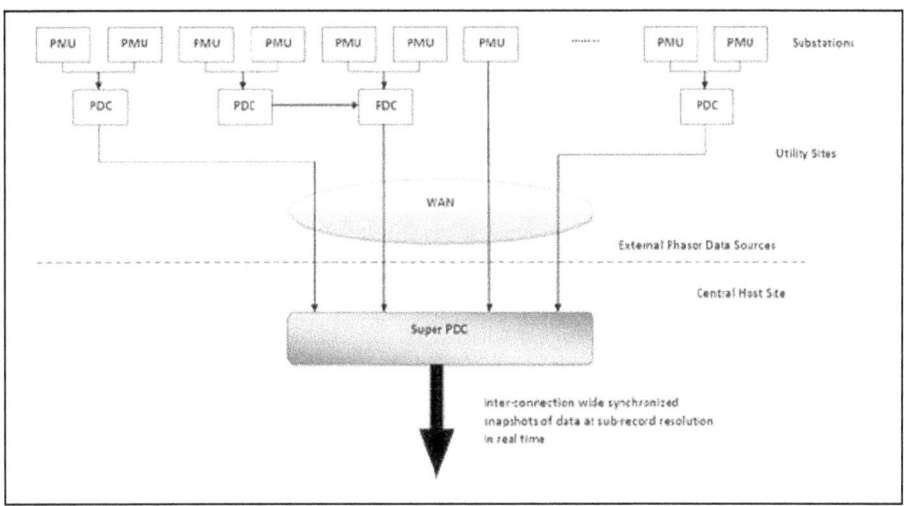

Figure 1.4 WAM System Based on PMU [24]

For transmitting the data of phasor measurement unit, the selection of communication channel forms a significant area of consideration. The data of the PDC requires an extensive bandwidth for the process of communication while the data from an individual PMU can be sent with the help of a precise bandwidth of communication-based channel [24].

Use of PMUs for making grid smarter

PMU is a significant innovation which plays a key for developing an intelligent smart grid. This innovation of WAM might be utilized for the below mentioned features :

a. Blackouts Threat and Scope Reduction

Fundamentally, power outages commonly known as blackouts are activated on account of rare events that are low in probability or are due to various uncorrelated or uncontrolled

INTRODUCTION

occasions. For prevention of these blackouts or power outages, well chalked out plans are important so as to control and protect the framework in crisis.. The PMU helps to build the protection systems .Thus, minimizing the possibility of a blackout state [19]. PMU helps in providing system information at a precise time instant . This information provided by PMU is used for dynamic real time analysis , which is further utilized for prediction of pattern-based frequency, power profile-based active-reactive etc. Security indices and margins will be identified and calculated by using real time information. This information is used for the prediction of emergency states, premier system security on the basis of monitoring and detection, and initiates restorative form of actions in order to avoid the instability of the system [104].

Information given by PMU will be useful for the analysis of PS noise like disturbances usually gathered from records of information-based loggers situated in the system framework. The GPS data which is synchronized permits an overall understanding of string-based events .

b. State Based Measurement

The operational working of an electric power system (PS) can be smoothly obtained at any instant if along with the network-based topology, the complex form of voltage phasor at each of the system bus is also given . These complex voltage phasors help in specifying the system completely , therefore it is usually denoted as the state of the system or the system state [61]. The SE i.e. state estimation function uses telemetered type of estimations/measurements of power injected at the system-based buses, reactive and real powers, status of circuit breaker, and voltages at the generator bus and so on for generating an optimised state estimation (SE) of the system. For a consistent and reliable form of SE, it is commonly needed that the measurement numbers must be larger than state numbers. Such a condition of operation is known as the observability-based criterion [22]. Further from providing an optimized state estimation, the system estimator helps to filter out, search, and detect all the gross errors in the set of measurements also called bad data detection.

INTRODUCTION

In traditional form, the state estimator-based input measurements were usually provided by RTUs i.e. Remote Terminal Units at the substation premises. Such kind of measurements did not involve the phase-angles as these are associated with certain difficulties with the process of measurement-based synchronization. Subsequently, the concept of phase angle was measured with a reference known as slack bus reference. Moreover, with arrival of PMUs this complex issue can be reduced as the PMU helps in measuring the current and voltage phasors generally synchronized through the GPS. The traditional SEs use the ICCP i.e. Inter-control Centre Communications Protocol for collection of asynchronous type of data with a rate of sampling rate of about 1 sample per 4-10 seconds. The measurements of the system are corrupted by noise . The relation amongst state variables and measurements is of non-linear type which is to be linearized and hence such a solution to the optimized problem is generally attained through the various (iterative) techniques. Due to the non-linearity various adjustments are required in designing of estimators and algorithms. These conditions limit the traditional state estimator which is prone to issues of convergence, hence distressing their reliability and accuracy particularly when the system gets stressed. Though, if in case a PMU is located at each of the system bus, then the relationship amid state variables and measurements is of linear form and a least square non-iterative solution can be utilized for determining the system state.

Due to economic and technical constraints it may not be possible to install the phasor measurement units at each system bus. However, a system is still observable by placement of PMUs only at chosen buses. Consequently, the already existing SE forms can be easily enhanced by utilizing the data from certain installed PMUs at typical locations. The data from such kind of units may be promoted as pseudo kind of measurement in the traditional SE as depicted in figure.1.5. The experimental analysis of such kind of installations have usually conveyed various advantages such as increased system accuracy, increased estimator stability, increased redundancy, and less time-based computation, etc. [23].

INTRODUCTION

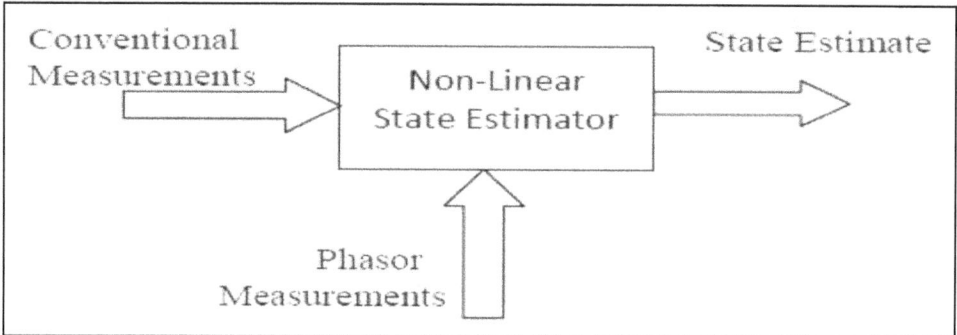

Figure 1.5 Hybrid form of State Estimation

A number of algorithms have also been stated for determination of optimized PMU locations in a system framework. Such methods usually depend upon a numerically-built topological formulation and use the optimized methods like integer-based programming tabu search, decomposition, binary search and particle swarm, etc. [9] [22].

c. Increasing the Transmission Line Capacity

The problem of transmission congestion involves a key important field for a secure and reliable power system operation. However, in a trendy de-regulated network of power system, its significance has been increased in multiple ways. The problem of transmission-based helps in limiting the activities of the market in a de-regulated market and lowers the potential benefit margin for the involved form of utilities. The line-base capacity of transmission is of limited form due to the following :

- Limiting Current - Governed by electric and thermal limits of a specific line.
- Transfer Capacity in Total- It is defined by certain area-based operational limits around the specific line.
- Transfer Available Capacity- It is defined by system operational limits. The strategy for congestion and real-time management is designed on Nominal Transfer Capability (NTC), generally calculated on off-line basis by utilizing traditional approach based on operational and environmental assumptions. Consequently, an inherent uncertainty exists with the limitations . Though, prudently additional

INTRODUCTION

margins of security are maintained which are normally lower than the genuine point of confinement. WAM technology gives a synchronized data estimation from densely populated areas, permits calculation based on transfer capability on real-time basis [25] that implies that the lines can typically be operated further near its points of limit, which subsequently expands the transmission system capability. It is also conceptualized that when the data of PMU is linked to other technologies of smart grid such as smart sensors that measures the sag, line temperature etc., then the limit of operation is further enhanced and leads to reduction of network-based congestion.

d. Calibration of Instrument Transformers

The operational monitoring and planning of a PS generally rests on the measurement of voltage and current signals obtained from instrument transformers secondary circuits. For an appropriate measurement, the secondary or auxiliary-based parameters must be in strict proportion to primary-based parameters. Though, this case does not belong to practical form of instrument transformers and the system errors such as phase angle error and ratio error are always present in the system. Hence for an accurate estimation or the measurement of the system, it is commonly required that the ITs to be calibrated appropriately[19]. The adjustment (calibration) of the system may be done by measurement scans for varying loads . Also for adjustments, good voltage transformers are required. It is additionally conceptualized that as the degree of PMU-based penetration in power framework increases , it will enhance reliability for ITs while disposing of a significant error source in the overall measurements [32].

e. Integration of Renewable Resources

The traditional grid has been modelled around so as to be able to dispatch the centralized generation. Predictable loads allow an open-loop control of grid power. Also , the environmental concerns have boosted the integration of various sources of renewable energy leading to cost reduction hence is promoted to form an integral section of smart grid. Indian economy supports great potentials of clean energy generation. By 2020, in India, the total

INTRODUCTION

capacity of renewable sources is projected to reach 8-8.5 GW that accounts for 11% of the entire power capacity. For an efficient and effective harnessing of this potential it is required to integrate sources of renewable energy with the technology of power grid. Though, such type of integration represents a big challenge for the operators of the system as both wind sources and solar photovoltaic are all inverter based whether they be connected to distribution / transmission systems. This will completely alter the system-based dynamics. SCADA systems used, provide a steady-state image which is very slow in longer time intervals, whereas the technology of synchro phasor in case of PS has created opportunities for an improved control and monitoring of the system for real-time management. Such kind of measurements can be generally utilized as snapshots system and consequently, shows the power system dynamics [23]. There is a significant portion of wind and solar power plants in the framework of power system hence, important decisions regarding the impacts on the bulk PS framework could be taken based on frequency values and voltage phasor quantities. The voltage phasors measurement at various bus-bars in PS framework and the interconnection corridors, and presenting them to the operators is a major step for alerting the operator about the status of the system [105].

1.2.3 Intelligent Electronic Devices (IED)

The basic idea that lies behind smart grid technology has led to an advanced technological evolution that helps in making the system smarter, more effective and efficient .The objectives of such technologies, mainly aims to locate critical challenges faced by electric grid systems on current basis, which branch largely from its aging infrastructure. In the past performance of power grid it is seen that the field of instrumentation over the grid quickly reaches to a life cycle limit, which negatively affects the global system efficiency and reliability. Hence, the conventional grid-based devices are not at all capable to handle the modern trends of power demands or the increased energy distributed sources or be able to handle the transformed standards and requirements of grid. Finally, as a result, there is no sustenance for modern upgrades and developments like electric vehicles or fluorescent lights, computers, etc. This is because the operation of grid is generally based on closed

INTRODUCTION

hardware, is vendor-defined and the software platforms make it effectively challenging to adapt as grid standards and requirements. Along these lines, a re-assessment of current grid framework is essential where the basic form of automation devices is taken to a more elevated level of intelligence to empower decentralized decision-making and distributed data acquisition. Another age of IEDs i.e. intelligent electronic devices is being evolved. These gadgets/devices are usually equipped with cutting edge innovations that make possible a two-way digital communication where every device on the system has capacities of sensing so as to be able to collect significant informational data about the grid Using remote automation and control, these gadgets can effectively balance and control at the node-based level as disturbances and changes on the grid framework happen. These IEDs usually communicate with SCADA framework and with each other for promoting distributed intelligence to accomplish quicker self-recuperating strategies and fault location.

At the core of these devices for the smart grid network lies the incredible innovation of the FPGA. Once observed as an innovation just accessible to engineers who have hands on experience of digital hardware, the advancements in this field have redefined the application of IED for smart grids [107]. FPGAs represent reprogrammable silicon chips that usually offer a similar adaptability for software programming running on a processor-based framework. In any case, because of their parallel nature, FPGAs are not constrained by the quantity of existing processor cores.

Also, they don't utilize the operating frameworks and limit the concerns of reliability with deterministic hardware and parallel execution to each and every task. Every task is independently allotted to a committed chip segment and can work in an autonomous way with no impact from other logical blocks. Thus, the performance of a single section/part of the application remains unaffected when extra processing is included.

INTRODUCTION

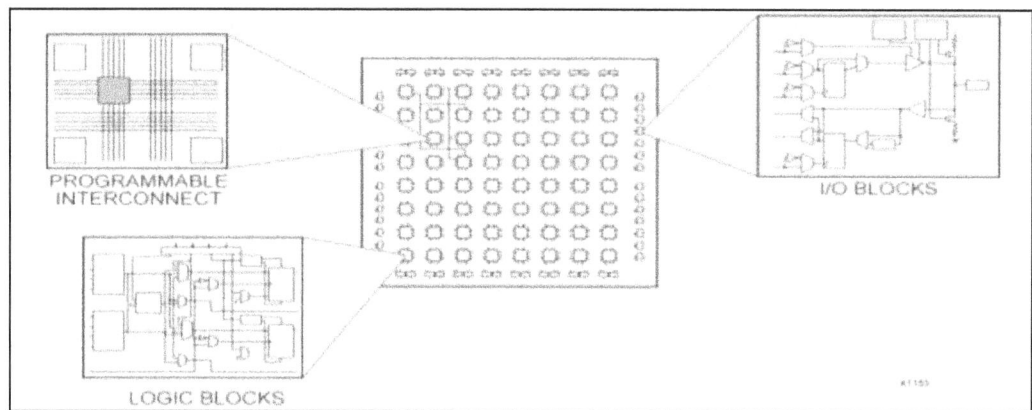

Figure 1.6 *FPGA circuit [107]*

FPGAs surpass the power of computing of PC processors and advanced digital signal processors (DSPs) by dividing sequential execution paradigm and achieving more per clock cycle. With the capacity to control both input and output hardware, FPGAs are able to achieve specialized functionality and faster response time to coordinate application necessities. Moreover, FPGA controls the installed instrumentation and control frameworks for the most recent age of IEDs over the smart grids, yielding extra reliability and flexibility, which empowers assembly of numerous multiple functional into a solitary unit, bringing down the expense of smart grid network frameworks (in whole). As FPGAs are consolidated into virtual instrumentation stages, this presents a key move from conventional hardware driven instrumentation frameworks to software (programming) driven frameworks that investigate the power of computing, display, connectivity, connection capabilities, and productivity of well-known workstations and personal computers. The platform for virtual instrumentation uses FPGA innovation, for example, National Instruments Compact RIO hardware, can join future adjustments to keep pace with grid network prerequisites that are ceaselessly changing [27]. Along these lines, as IEDs for the smart grid matures, functional developments of modular hardware and open software can be utilized instead of modifying the board design or replacing the device.

INTRODUCTION

1.2.4 Two-way Communication Technologies

Using the technologies of smart grid that help in improving the communications between power providers and their customers represents a significant key to have distribution systems running more reliably and efficiently. There has always been a rising trend within the industry of utility that is focussed or centred on improving the basic knowledge of the companies for gaining an improved grid infrastructure. The number of consumers that seek the service of energy is increasing day by day, but the infrastructures supporting both the transmission and distribution of such power are not quickly expanding. However, most of the utilities highlight improving communication and control along the grid technology so that they can distribute their power more efficiently, and in doing such practices, the demands of the customer are fulfilled.

Advanced Metering Infrastructure or AMI usage: Most of the companies have eventually started using the capabilities of AMI and smart meters in order to build the knowledge over the smart grid technology. The Smart meters contain large number of capabilities that provide help both to the service provider and the customer. For example, they can easily send and receive the signals on the basis of consumed on-site power, such as the office or home, and it may communicate the power back to the place of utility. Such type of communications can assure reliable form of electricity [30] [40]. The energy service provider can also use the above information to have better knowledge and understanding of electrical loads flowing through their provided infrastructures. In performing this, they can more impressively and accurately account for large number of times when the demand is high, and may even mitigate the power blackout possibility.

CIS Software and Advanced Billing: The process of billing and satisfaction of the customer could easily attract the market scenario. Such type of systems can smoothly allow the utilities to design more models of dynamic pricing for their users, which in return help in power conservation at certain times of the peak demand.

INTRODUCTION

Specifically, the consumers are charged in a format on the basis of static meter-to-cash format, however, such type of systems could allow the billing operation to become more focused towards the users on the basis of Navigant Research Report Study. Such a technology could help a much better integration of new resources of power that enters the grid, involving renewable energy and natural gas. As a result, CIS software and billing services are generally expected to rise from $2.5 billion (2013) to $5.5 billion (2020), in accordance to the figures from an institutional firm.

Energy Storage and Power Electronics: This includes High Voltage DC i.e. HVDC, FACTS in order to allow integration of renewable energy sources and distance transport along with other devices of FACTS like series capacitors and UPFC i.e. Unified Power Flow Controllers. Since the electric power was first designed about a hundred years back, it has been individuals' need for living. As indicated by economics of business constrain for electrical energy, the power must be immediately used following generation. However, with smart concept of grid technology, the distributed form of generation becomes more encouraging, while the options for energy storage have been discovered to be feasible. Presently, the distributed Energy Storage Systems (ESS) have become very impressive. Table.1.1 represents a service of energy storage system for an electric power grid in all the levels of power system that involves generation, transmission, distribution and the end users. Fig.1.7 represents a block diagram covering the significant components of PEBES i.e. power electronics-based energy storage. The circuits of Power electronics play a significant role in regulation of raw energy from distinct devices of energy storage and grid interfacing operation. Very fast response and high efficiency of power circuits of electronics makes it very impressive and attractive in an environment of smart grid.

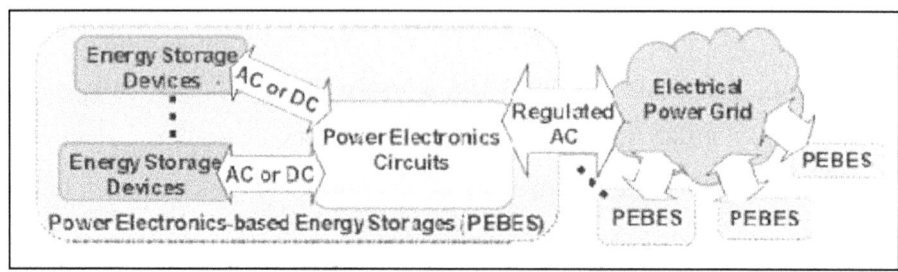

Figure 1.7 Power electronics based energy systems. [104]

INTRODUCTION

Table.1.2 Energy Storage System Services

	Power Application	**Energy Application**
Generation & Transmission	• Damping Power Oscillation • Voltage Support • Supplementary Reserve • Black Start	• Energy Time-Shift • Process of Transmission • Congestion Relief • Deferral for Transmission Upgrade • Electric Supply Capacity
Distribution	• Quality of Power • Voltage Support	• Reliability of power • Intermittent type of Mitigation • Distribution Upgrade Deferral
End Users	• Power Quality • Management of Demand Charge	• Interruption Backup • Energy Time-Shift

The devices of energy storage, for instance, are flywheel, mechanical generators, battery, fuel cell, generally driven by the prime mover, etc.

Communication

Inside the environment of Smart Grid, real-time and reliable information is a complex aspect to assure a reliable power distribution from the unit of generation to the unit of end user. Henceforth, control and monitoring are required to be dealt in an intelligent way, empowered by communication technologies and modern information, this represents a critical need or requirement to assure efficient and effective management and operation of the system [10] [20]. Further assuming that the utility contains a wide-ranging backhaul network in place, so there is an additional need for AMI communications which will involve facilities of online based communication and the smart meters. Usually two communicational structures are needed to get the informational flows within a system of Smart-Energy Grid:

- From electrical appliances and sensors to the smart energy meters
- Between the data centre utility and the smart meters.

Many of the distinct type of networking technologies and communications, using two basic mediums of communication i.e. wired and the wireless are usually available for the support of

INTRODUCTION

distinct applications of the Smart Grid. These involve conventional phone lines of twisted-copper (ADSL and DSL), WiMAX, microwave, fibre optic cable (OPGW etc.), power line carrier, cable lines, cellular (GPRS and GSM) satellite, and the service of broadband over power line along with the short-range technology in-homes like ZigBee and WiFi. From the statements it is confirmed that, regardless of networking innovations employed, the wireless-based communications play a significant part in the Smart Energy Grid communicational deployment and it will further need an additional important spectrum of radio-based frequency [31]. Hence, for accommodation of spectrum needs for the environment of Energy based Smart Grid, it will definitely be cautious for system utility to involve system regulator for exploring certain possible forms of substitutes; this may involve an additional process of spectrum leasing, or it may even form the option for the purpose of inspection in order to perform spectrum sharing process with other type of users. Moreover, prior to such compromises occurring in the system, it will be prudent for the utilities of the system to conduct several surveys, in partnership with effective service provider of telecommunication, to perfectly observe the spectrum requirements and also it obtains an impressive insight in regard to specific uses, for example e.g. fixed and/or mobile access.

Table 1.3 Technologies of Smart Grid Communication

Technologies	Rate of Data	Spectrum	Coverage	Applications	Limitations
GSM	Up to 14.4 Kb/s	900-1800 Mhz	1-10 km	HAN, Demand Response, AMI	Low Rate of Data
GPRS	Up to 170 Kb/s	900-1800 Mhz	1-10 km	HAN, Demand Response, AMI	Low Rate of Data
3G	384Kb/s-2 Mb/s	1.92-1.98 Ghz 2.11-2.17 Ghz	1-10 km	HAN, Demand Response, AMI	Expensive Spectrum
WiMax	Up to 75 Mb/s	2.5 Ghz, 3.5 Ghz, 5.8 Ghz	10-50 km (LOS) 1-5 km (NLOS)	HAN, Demand Response, AMI	Not Extensive
PLC	2-3 Mb/s	1-30 Mhz	1-3 km	AMI, Fraud Detection	Severe Noisy Environment
Zig Bee	250 Kb/s	2.4 Ghz 868-915 Mhz	30-50 m	HAN, AMI	Low Data Range, Limited Range

INTRODUCTION

1.3 Wide Area Monitoring

In the present typical scenario, the concept of wide area monitoring systems popularly known as WAMS, the measurements in a synchronized manner are generally gained from phasor measurement units and each and every form of data is usually sent through the network communication, concentrated and received at a control and decision supporting system known as PDC i.e. phasor data concentrators that help in determining accurate corrective, protective, and preventive measures. The decisions that help in determining the system support will help the operators at the controlling centers in order to take intelligent and smart controlling actions of the operator. Such active actions are further converted into the signal-based feedback sent through the network of communication in order to exploit the protection and controllability of the power resources in a planned framework of power system. PDC and PMU are thus considered as the backbones of WAMS architecture. PMU represents a logical or function device providing angle and magnitude-based synchro phasor, frequency of the system and changed rate of frequency-based measurements on the basis of collected data from more than one primary sensor such as potential (PTs) and current (CTs) transformers. PMUs may provide an optional information like evaluated reactive (MVAr) and real (MW) powers, Boolean status words, and sampled measurements [1] [13] [33]. PDC, a logical or function device, generally operates as a communicational node in network where data of the synchro-phasor from a finite number of PDC and/or PMUs s is gathered, aggregated, aligned by time, and usually sent out as an individual type of stream to the higher applications of PDC level . The concept of PDC is to implement real-time control applications and wide-area protection on optional basis [3] [6].

With rising number of installed phasor measurement units in the wide area systems , it requires an active and effective architectural data management and collection.

<u>Definition and Classification WAMS Architecture</u>

The architecture of WAMS is generally classified as Distributed, Centralized, and Decentralized architecture [30]. The factors distinguishing among such type of dataflow or information between data acquisition location are the selective location and the place where

INTRODUCTION

the actions are based on certain decisions performed. The section below describes distinct types of wide area monitoring systems i.e. WAMS architecture

WAMS Centralized Architecture : In an architecture of WAMS (centralized), the PMU-based data acquisition, and performance of therapeutic activity is generally performed at the central type of location. Figure 1.8 discusses the architecture of WAMS (centralized). The phasor monitoring units (PMUs) from certain operating substations forward the phasor data to the PDSC Central part where concentration of data and the time alignment of all the received PMUs activity of the data takes place. The data concentrated is generally used for the process of visualization and analytics. The corrective actions derived from such type of analysis is forwarded to the primary devices.

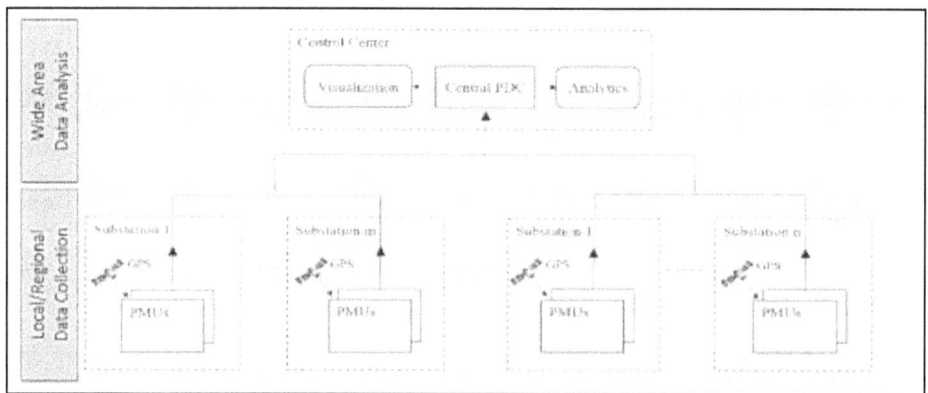

Figure 1.8 Centralized WAMS Architecture [30]

WAMS Decentralized Architecture : In an architecture of WAMS (decentralized), the area of monitoring is generally divided into small multiple type of areas and the PDCs perform local control of such small regions locally . The system controllers built locally are linked to each other if they have the capability or the requirement to resolve the problems of larger area. Fig.1.9 encapsulates the decentralized architecture of WAMS. The local area PMUs like a specific region or substation send the phasor data to the local PDC respectively for the purpose of system operation or processing. The local type of PDCs help in data analyzing process in order to perform any kind of remedial action for protection of their respective local assets. Though all the local distributed PDCs that are associated to one another for

INTRODUCTION

exchanging data for the purpose of monitoring and controlling large area, but this is not an efficient possible solution for wide area monitoring. The concentrated coordinated acquisition of data from the local type of PDCs and their operational analysis for wide area monitoring is often critical and challenging and it does not reach its target in most of the times.

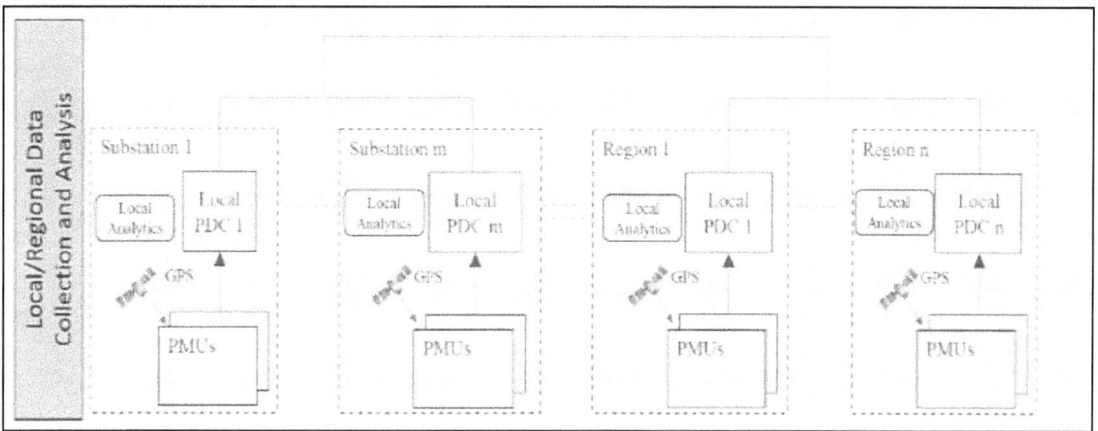

Figure.1.9 Decentralized WAMS Architecture [30]

<u>WAMS Distributed Architecture:</u> The architecture of WAMS (distributed) can be usually mapped between the architectures of decentralized and centralized form. It involves local as well as the central type of controllers. It is depicted as a centralized type of control with its decentralized form of running stage. Fig.1.10 discusses the architecture of distributed type of WAMs. It is usually comprised of PDC (local) placed at regional level or substation and the master form of PDC placed at central controlling station. The PMUs present in local region like a specific region or substation send data phasor to local PDC respectively. The overall local PDCs present are linked to the master form of PDCs at central controlling station. The major difference is present in the information flow, as the local PDCs may perform processing of the data PMU in a local way, controlled and supervised by PDC master.

The architecture of Decentralized texture is not considered for the purpose of analysis owing to low implementing popularity and the other drawbacks of the architecture mentioned above. The two considered architectures have been used and analyzed on the basis of mentioned WAMS elements that are successful or significant in WAMS deployment such as

INTRODUCTION

infrastructure, vulnerability of data, communication, sharing of data with external objects and entities and availability of the system among others.

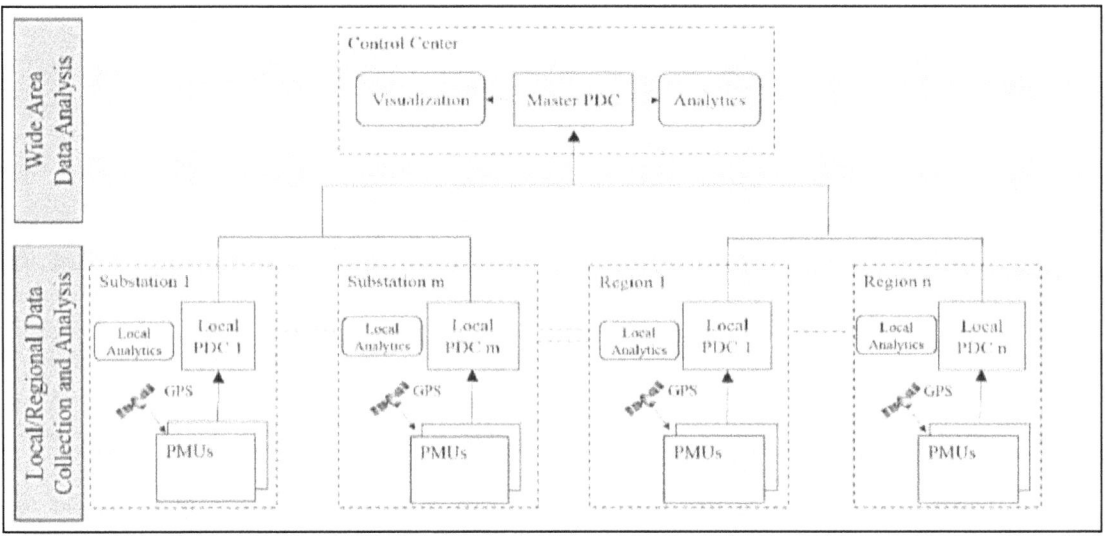

Figure.1.10 Distributed WAMS Architecture [30]

Communication of data: The process of communication in the architecture of WAMS can be usually characterized by three elements, the bandwidth, infrastructure, and latency for the process of communication. The section below generally deals with every aspect and describes how the distributed and centralized architectures fare and stand against each other on the basis of different aspects.

a. Bandwidth: In WAMS architecture, all of the PMUs send data phasor on direct basis to the central type PDC for proper concentration. In case of distributed architecture of WAMS, all the phasor measurement units send the phasor data to the substation or local PDC respectively. The local form of PDC provides the consolidation through data aggregation and the process of time alignment. This aggregated form of individual output data stream from substation PDC and is sent forward to the master PDC at the controlling station. The local PDC reduces the communicational bandwidth needed for the data that is to be forwarded from the substation to controlling station.

INTRODUCTION

b. Latency: The main aim of latency in the architecture WAMS is generally characterized by the PDC based latency and communication-based latency. The latency of PDC is further comprised of device latency PDC and the wait-time of PDC. In an architecture of WAMS (centralized), all of the PMUs are connected directly to the central type of PDC and further they usually tend to have less communication-based latency. In such an architecture (distributed), a local form of PDC or substation PDCs introduces a few milliseconds device-based latency. For the waiting-time latency, each of the PDC sticks to stay and wait for few of the user-configurable time durations such that slow form of phase measurement unit (OMUs) data can be processed for time-based aligning operation at region of PDC. The architecture of WAMS (distributed) having PDCs based hierarchy does not involve addition of latency-based waiting-time though with an architecture of centralized, the central type of PDC generally has to wait and stay for operational alignment on the basis of time.

c. Communicational Infrastructure: The architecture of WAMS (centralized) needs most of the communication links same as that of PMUs linked to PDC i.e. phasor data concentrators in order to forward the phasor-data to central PDC on direct basis. The number usually doubles itself when a specific centralized form of WAMS deployment needs duplicity in the infrastructure of communication. An architecture of WAMS usually needs an individual link of infrastructure-based communication as the substations forward single or individual stream of data to the control station

d. Data Storage: The architecture of WAMS usually needs a PDC for storing the phasor data for certain applications like data forward to high PDC levels and the data for its function-based analytics.

Data Security: In an architecture of WAMS in its distributed form, all the PMU-based data is locally concentrated and only individual stream of data is forwarded from the substation to the controlling station. The two major security layers of communication are designed inside the substation area with the help of existing measures of the security for the overall communicational data and the other kind of secured and safe means like individual stream of data encryption outside the substation area. This usually boosts the system and data security

INTRODUCTION

as compared to architecture of centralized type where all kinds of PMUs use data multiple channels for sending direct data to controlling station.

Event and Alarm Management: The centre for control conducts and performs stability-based analysis on the basis of possible/available form of data phasor using control, protection, and monitoring applications. Such an analysis helps in identifying the abnormalities of the grid operation. It usually involves the generation of various kind of alarms and performs various events for its fast-remedial actions that have to be performed on certain grid levels. The coordinated form of event management and alarms placement is easier in case of architecture with centralized structure as compared to the one with distributed structure as it does not involve large action levels to traverse through it. The architecture in distributed form includes PDCs hierarchy and hence the multiple levels that make it really difficult to properly manage the events and alarms.

Cost of implementation: The architecture of WAMS (centralized) provides a warrant to multiple communication links and central PDC in order to connect or associate with all of the PMUs whereas in case of architecture of distributed forms generally require a single communication link and multiple local PDCs connecting to the control centre. This clearly indicates that in case of distributed WAMS architecture, the cost of the implementation is more.

1.4 Smart Grid Distribution

Distribution intelligence essentially denotes the section of Smart Grid i.e. applicable to the utility-based distribution system such as the transformers, wires, and switches that brings the utility-based substation to the customers. The power lines passing through the back yards of people are considered as one of the parts of electrical power distribution system (PDS). A significant component of distribution-based intelligence is the response and the outage detection. Presently, most of the utilities generally rely on the calls of the consumer in order to know the affected areas of distribution system resulted by a power-outage [1] [11] [22]. With the help of smart meters, distribution-based intelligence will surely assist to rapidly point out the sources of power outage so that maintenance teams can dispatch immediate

INTRODUCTION

services to the problem creating area. Considering this, the utility-outage response can be easily improved. Most of the utilities count on manual switching and power distribution complex schemes in order to continue the flow of power to most of the consumers, even when the power lines are destroyed and damaged. Though, such a method has its own drawbacks, and in most of the cases a system that is of automated form responds faster and could continue the power flow to more consumers. In order to determine the best solution to power outage, the system requires sensors joined by intelligent system with an automated switching [105]. The power of the system can be re-routed to most of its consumers in a matter of few seconds, or even possibly in milliseconds. It is also probable to quickly respond to disturbances such that only the surrounding neighborhood gets affected instead of all, while other sources of power are re-routed rapidly to avoid any power interruption.

Along with system-based outage response and detection, another potentially distributed intelligence application represents the ability to provide balanced optimization between reactive and real power [109][110]. The devices that release and store energy like the system-based capacitors or wire coils in order to create the magnetic fields same as the electrical motors, having the ability to generate large electric currents without consumption of real power, which is usually called as the reactive power. Only a certain amount of reactive power is of desirable form within an electric power system, but large amount of reactive power results in large flows of current, serving no purpose, resulting in various efficiency losses as they usually heat up the wires of distribution system. An intelligent system of distribution uses the power-electronics facility for maintenance of proper reactive power level in the System. An intelligent distribution system supports control and protection of the power lines and the feeder lines that contribute in designing of the distribution system. Most of the lines of feeder are presently protected by relays or breakers that provide a mechanism of tripping when a heavy current flows in the power lines, it arises from a situation caused by the system fault. Sometimes such relays incorporate and develop time delays in order to allow high current momentary flows that can be caused by power equipment's of the industry, then that of a fault. The Protection systems represents the joint effort of instantaneous breakers along with settings of relays or time-delayed breakers with lower

INTRODUCTION

amount of settings. Such relays and automated breakers systems ends up providing a balanced act: they should permit the system to run with large currents whenever required. But at the same time, it must provide protection to the system itself and the humans surrounding the system from large flows of heavy current. The distribution intelligence is capable to provide an elegant strategy for protection of feeder lines with the help of using practical control and monitoring to first observe and detect and further correct the faults by maintenance of system-based reliability in case of non-faulty operative conditions [26]. The intelligent systems could easily isolate and detect the faults in unique equipment and direct the power path through backup system while sustaining the reliability. Intelligent distribution also helps in incorporating more practical detectors for the ground-fault in order to minimize the possibility of shock to the people dealing with the power lines.

1.5 Conclusion

The innovations for a smart grid have become significant. As innovation developed, frameworks and devices which can strengthen the procedure of an increasingly more intelligent grid came into existence. The concrete energy-based policies encourage Smart Grid activities around the world. The practices of smart grid in numerous areas barely display competition but, in its place, an uncontrolled network of comparative desires and common exercises is displayed. This chapter out of the need explains the development of Smart Grid to modernize the framework of electric network. In all these years the conventional framework became stretched and required more features. It has highlighted the characteristics and functionalities of Smart Grids. It has displayed the Smart Grid basics and related innovations and has discussed the exploration exercises, issues and difficulties that rotated around them. There are oppurtunities for research in the field of Smart Grids for power quality and reliability studies, cloud computing, battery frameworks, and practical significant sustainable power sources integration. The practical feasibility of the Smart Grid is not encouraging while relying on the intricate present framework. Even though there are technologies and experiences available for reference, the Smart Grid (SG) represents assumption of money, time, and continuous testing and examination. The attempts are set forth for Smart Grid as this helps in accomplishing ecological protection, safeguarding, and

INTRODUCTION

energy maintainability. The final destiny of the SG may be hard to anticipate, yet continuing developments display a promising future with amalgamated technologies.

Chapter 2 LITERATURE SURVEY

2.1 Introduction

The term "smart grid" represents the amalgamation of various technologies in order to enhance the grid-based power efficiency. This introduces new controls, sensors, and high speed of communication for adjustment of grid parameters in dynamic form. Most of the methods provide a bulk use of efficient sources of generation in order to improve the present 35 percent average mark.

The main objective of smart-grid is to manage the loads dynamically. Most of the system complexity as well as the managing cost of grid arises from the demand-based variability. The complexity driven by such a demand curve is presented in Figure 2.1 given below. For most of the region, the grid-based electrical energy is not able to effectively perform the process of storing the energy that means that the production should chase demand. In order to account for such demands of variable type, the utility-based companies mostly use base-load type power plant for achieving a stable demand supplementing the peak-load based power plants [14] [30]. The base-load type of plants is specifically nuclear or hydroelectric, coal-burning. Such base-load plants carry high cost for modelling but are not expensive in their operation that means that they are of cost-effective nature if they run on constant basis and at their full system-based capacity. Additionally, they mostly have a long time for start-up and are usually not constructed to frequently turn on and off, hence they are not good for dynamic peak demand .

On the other hand, the peak-load plants are capable of quickly turning on during peak demand times. The peak-load plants are not very costly for the build-up process, but when these plants are compared to the base load type of plants, then these get more expensive during their operation, and thus these are most useful economically when they add more inexpensive supply to the base-load. For such kind of purposes, gas turbines are used. The process also includes the intermediator plants having attributes lying between peak-and base - load plants.

From such a methodology, one can observe that it is basically more effective to have more even form of demand curve, so that energy obtained from base-load plants is cheap and can be utilized more evenly. Thus, somehow it is needed to flatten the demand curve which is known as "load levelling". For doing this, ultimately the utility-based companies should provide certain benefits for the consumers in order to lower the use during the peak demand. For providing such kind of incentive, the utilities begin to roll out "smart" meters that are to be used by the customers having variable type load that changes throughout the day instead of billing the overall consumption at a single flat rate. During the peak hours when the prices are high, it encourages the consumers to transfer the consumption of energy to off-peak hours, thus it flattens the demand-curve, granting for less costly global production of energy and lower total prices (ideally).

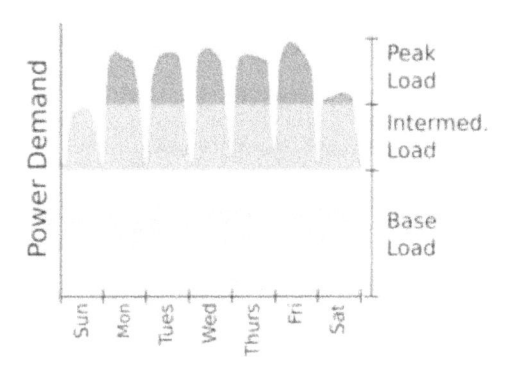

Figure 2.1 Characteristically shaped demand curve, viewing different levels of power demand [110]

Additionally, for encouraging the consumers to reduce the usage manually during the peak hours, a smart-grid grants help to large appliances in an automatic way like clothes dryers and water heaters to reduce the cost and progress the grid-based stability. For example, detecting the reduction in grid's frequency which represents the production of grid not matched by its consumption, the appliance sheds its load power in order to preserve the grid stability. The Pacific Northwest National Lab experimentally performed a trial run where the cloth dryers' clothes were computed to switch off their heating elements under distinct type of situations. In spite of this, the disturbances of the system were low enough in a particular duration of time, such that the applicants have never observed the difference, which indicates the promising nature of the methodology used. Perhaps, consumers who choose the usage of such kind of controlling devices could be rewarded or compensated by the utility-based company with a discount or a credit.

In brief, a good stress deal is located over the sector of electrical power generation in order to provide for a high fluctuating demand. Most of the aspects of a global approach of "smart"

LITERATURE SURVEY

grid, includes load-control switches and variable kind of metering, having a great strength to ease such kind of stress that allows more efficient and effective global electricity consumption as well as the production.

2.1.1 Demand Based Response

The concept of demand-response represents an approach of performing a large control over the global system of electrical grid by the use and availability of individual or single system facilities' in order to lower the system usage, if the grid gets extremely stressed. With the flexibility for energy management and impressive two-way communication, in addition to storage capacity and onsite generation, buildings are prepared well to grab the benefit of Smart Grid system and the programs-based on demand response. The players of demand response programs are usually offered with the following:

- Utility-based companies.
- Independent System Operators: ISOs which derive a region-based power pool or grid transmission over a single or more than one state. Their users/consumers take part in programs of demand-based response through distinct members known as curtailment service providers (CSPs) or aggregators, who play the role of agents for the consumers.
- The third-party CSPs or aggregators that makes a contract with ISOs or utilities. The global providers of institutional, industrial, and commercial demand-based response present the third-party aggregators. They generally work with most of the ISOs or utility-based companies that help the customers to design energy efficient and demand response programs.

2.1.2 Incentives for Demand Response

The concept of demand response is same as for instance, an airline system that pays the passengers to take the advantage of getting an off in case of over-booked flight; In the same way the consumers taking part in the grid system are paid to have an off over an over-booked system of grid. The owners of the system that takes part in the programs of demand-based response may see lowering of utility-based bills or they get other form of incentives from the administrators of the program. On contractual basis, the owners of the building might get a credit over the bill, or get reduced rates of kilowatt-hour for the global power.

LITERATURE SURVEY

The demand-based response events of the system are generally prompted by the peak-based capacity (i.e. reliability concerns) or the wholesale-based high prices. The program based on demand response is basically categorized as either time-based or incentive-based.

a. Time-based demand response: It consists of the consumers that make spontaneous reductions on the basis of price-based structures i.e. dynamic, real-time pricing or critical peak-based pricing at each of the given point on the basis of time. A reduction call is not started by controlling center that forwards a connection to consumers; it is usually more than a program i.e. market-driven, allowing the principles of demand and supply and the comparable prices to simply adjust the electricity demand. Such type of programs help to provide motivation to the payers of rate for load curtailment by passing along the energy-based wholesale prices. As the process does not involve an earlier or pre-commitment for participation, so the selective decisions on how the load shedding process is made are done only on the daily basis and building/model holders are not penalized directly for inactiveness, though the process of buying the electricity at peak price could be thought of as a penalty. When the prices are high, the energy spent gets lowered by cutting the consumer-based consumption, motivating the owners of the building. However, the demand response that is time-based generally needs an automated strategy of control to effectively function for the owners of the building.

b. Demand response on the basis of incentives: It represents a proper program, where a user of the system makes a contract for the participation purpose i.e. based on control center communication to the consumer for signaling each of the distinct type of event taking place. It consists of capacity-based or reliability-based programs, buy-back, demand bidding type of economic voluntary programs in order to provide generation supplement. The programs that are based over the system capacity are currently considered to be of most dominant type of demand-based response. The owners of the building are generally offered constant monthly fee by administrator of the program in transaction for the agreement to cut the demand by a possible amount at each of the event based on demand response. In addition, on per event basis, a small credit on the original load reduction gets delivered.

c. Receiving the demand-based response signals: Once the assessment of load curtailment ability is done, various methods have been planned and developed for the purpose of implementation, managers and the building owners must be able to find how they will execute and receive the signals of demand response.

- Demand response of manual type: It consists of personnel roaming through a provided facility to switch off the irrelevant lights. Due to this, it is considered as an approach that is of most labor-intensive type, it can restrict the process of load shedding at the time of an event.
- Demand response of semi-automated form: It consists of the operator of the building that begins a strategy of load shedding generally pre-programmed in a centralized form of control system.
- Demand response of fully automated form: It does not need the human-based interference. The controlled system in its automatic way executes a control strategy i.e. pre-developed on the basis of receipt of an externally linked communications signal, unless the operator of the building chooses to nullify it.

2.1.3 Strategies for Demand Response: The program participants regardless of the program must be able to evaluate when, how, and where the participants facilities use the power and it must find their capability of load curtailment. The main objective is to overcome the demand in its least invasive manner while forwarding the best service in return. The strategies of load curtailment may consist of the following:

- Shifting: Movement of demand from peak time to low-tariff time.
- Shedding: Restricting or limiting the load on temporary basis.
- Storage: For consumption at the demand or peak pricing, a thermal storage device or a battery bank collects the power (energy) during the off-hours.

Generation onsite: Using the generators' back-up and renewable energy instead of grid generated electricity during peak prices or demands.

2.1.4 Domestic Load Classification

The domestic load is generally categorized into three of the sub-loads i.e. the basic, regular, and burst type of loads. Fan, Lighting, Internet, TVs represent the base-line loads. Such type of loads operate at any time according to the basic need for its use and it does not represent the main part of load. The load of regular type is generally comprised of cooling, refrigeration as well as the heating loads that tends to be varied with conditions of weather. It basically runs for a large time period and is nearly around 43% of the domestic (total) load. On the other hand, the burst loads are assumed to run for a particular time period and these can be easily tuned on the basis of day to day routine such as heating, washing, cleaning, cooking, and ironing. The base line load is not the main issue that creates the factor for comparison with the burst as well as the regular type of load. So, providing first preference to base-line load will help the system to run smoothly [26]. For the heating systems and energy cooling, necessary steps must be taken to derive the load of regular type by using distinct energy efficient devices as it will increase the efficiency of the system by saving around 50% of the electric domestic power in total.

With the process of integrating distinct techniques of ICT i.e. Information and communications technology, the techniques of demand-side load management contribute significantly in improvement on the basis of power efficiency. It is generally estimated that the CO2 emission from 53 million cars would get replaced if the power of the grid is about 5% (least) more effective and efficient than the earlier used systems. The energy efficiency on the basis of demand side is considered to be more profitable. The system-based utilities have no control over the customers' load, the system utility helps in performing the demand-side management of load for planning, monitoring, and implementing distinct techniques for a smooth running of the overall grid-based operation[32].

The table below represents the general form of average power consumption taken on random basis.

Table 2.1 Average consumption of Power for distinct type of loads[26]

Load Type	Items	Watts	Summer Months Hrs/Day	Summer Months Days/Wk	Summer Months WH/Day	Winter Months Hrs/Day	Winter Months Days/Wk	Winter Months WH/Day
Basic Load	Fans	400	10	7	4000	0	0	0
	Lights	200	6	7	1200	8	7	1600
	TV	75	4	7	300	5	7	375
	Router	15	24	7	360	24	7	360
	PCs	80	4	5	228	5	5	285
Regular Load	AC	1800	5	4	5142	0	0	0
	Fridge	480	7	7	3360	3.5	7	1680
Burst Load	Washer	320	1.5	2	137	1.5	1	68
	Cleaner	2000	0.5	2	285	0.5	1	142
	Pump	1600	1.5	7	2400	1	7	1600
	Iron	1000	0.25	5	178	0.25	5	178
Total					17592 WH/Day			6290 WH/Day

LITERATURE SURVEY

The consumption of load gets accelerated in summer for the load of regular type, as it consists of more usage of power for refrigration and cooling units. The basic load is low in wattage but the consumption of power is usually high. It comprises of about 35% of power as its operational time is maximum when compared to other type of load as shown in figure 2.2 . If one can manage this type of load, then half of the issue could be solved easily for blackout in total.

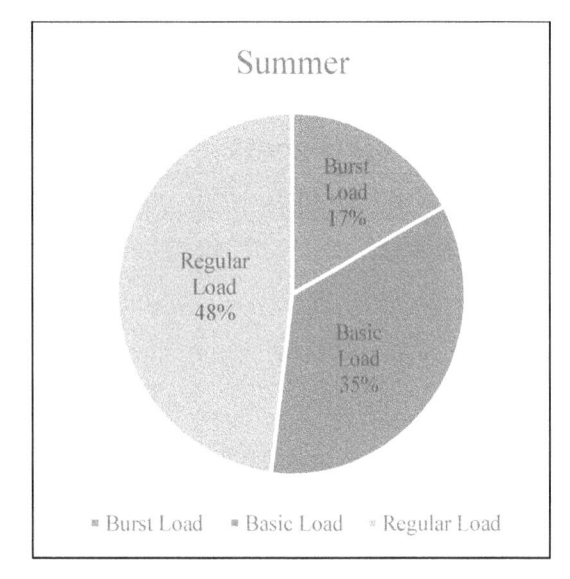

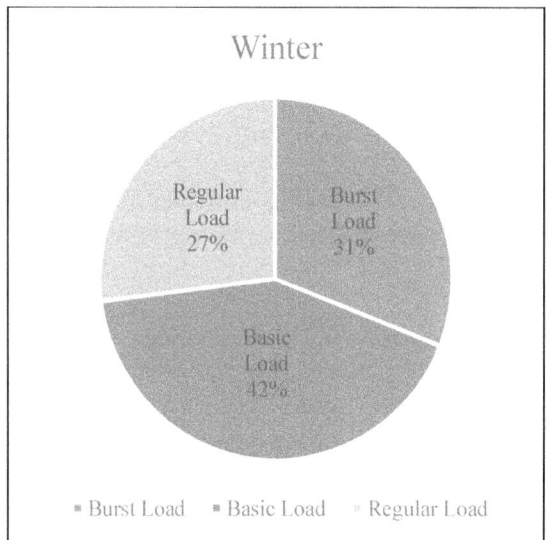

Figure 2.2 Load consumption comparison[26]

2.2 Smart Grid

Smart Grid is a revolution with respect to application of digital systems to electric power network[1]. It helps to offer large technologies that are not only valuable but can also be used for future or is already in current use. The technology of Smart Grid generally involves, digital control appliance, intelligent system of monitoring, and electric network. All of such methods, help to deliver the power from producers (owners) to consumers (end-users), controls the flow of energy, and makes the electrical network performance more controllable and reliable. In short-term analysis process, a smarter grid functions to work more effectively and efficiently, enables the system to handover the service level that an individual comes to expect in more affordable form in the era of boosting costs, while it also offers a significant amount of social advantage which involves the low impact over the environment. For longer term, one can predict the SG to bring about a massive change in our lifestyles .

LITERATURE SURVEY

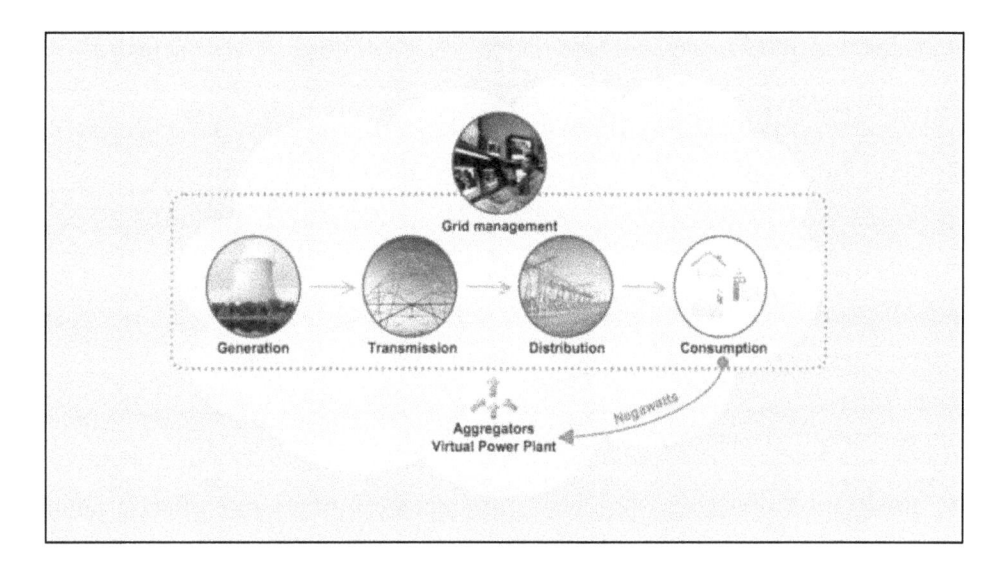

Figure 2.3 The Smart Grid adds intelligent, integrated communication and full transparency to the grid [70].

What makes a Grid "Smart?"

In brief, the digital technological development helps in allowing a communication (2-way) between the customers and the utility, and the power of sensing along the line of transmission forms the grid to be smart. In case of Web, the Smart Grid generally consists of computers, controls, new technologies, and automation along with all the equipment operating in synchronization. But in such a case, these technologies functions[9] with the grid in order to answer in a digital manner for rapidly changing demand of electric system.

What does a Smart Grid do?

The technology of Smart Grid helps to characterize an opportunity i.e. uncommon to derive the industry of energy into an advanced generation of efficiency, reliability, and availability that will surely dominate the environmental and economic conditions. In the period of transition, it will be very complex to dissipate technological improvements, testing, education of consumer, development of regulation and standards, and the sharing of information between various projects for ensuring the advantages that we foresee from the technology of Smart Grid to become a matter of truth. There are various advantages that are linked with the concept of Smart Grid including the following:

- More effective and efficient electricity transmission.
- Quick electricity restoration after the disturbances occurring due to power failure.

LITERATURE SURVEY

- Reduced cost of management and operations for utilities, and conclusively lowering the costs of power for customers or the end users.

- Reduced form of peak demand that will lessen the rates of electricity.

- Rising integration of renewable energy (large-scale) systems.

- Better customer-owner integration of energy-based generating systems, consisting of large systems of renewable energy [12][13]

- Improved form of security

Presently, disruption of electricity like blackout may result in domino effect-a failure series that creates impact over communications, security, banking, and traffic. Such a specific winter threat, when the home-based owners can be left without any kind of heat process. A smart form of grid will add up flexibility of electricity-based power system and it makes it well prepared for addressing several cases of emergencies conditions like severe form of earthquakes, storms, terrorist attacks, and bulk solar flares. Due to interactive (two-way) capacity of the system, the concept of Smart Grid helps to allow rerouting in an automatic way when there is outage or failure of equipment. This will further help to lower the amount of system outages and minimizes their impacts. The technology of Smart Grid detects and isolates the power outages, prior to their becoming large-scale system blackouts. The new concept ensures the electricity resuming strategically and quickly after the operational emergency condition. Additionally, the technology of smart grid will offer a benefit to consumer-owned energy or power generators for the production of power, if it is not at all accessible from the system utilities. By joining resources of distributed generation, a community of the system could make the police department, health centre, phone system, grocery, and traffic lights continue the operations even in case of emergency conditions. Additionally, the concept of smart grid represents a method of addressing an aging infrastructure that requires to be replaced or enhanced. It presents a way for addressing the efficiency of energy, in order to bring a rising awareness to customers about the link between the usage of electricity and environment. And it also represents a method to present a security to energy system drawing large amounts of electricity i.e. home-based, more contrary to attacks and natural calamities[107]

[108][109][110].

LITERATURE SURVEY

How a Smart Grid works?

A complex smart grid characteristic helps in production of power through the sources of renewable energy to add the power within the smart grid. The power derived from renewables is distributed and stored as needed in peak demand times or when the outage of power occurs and hence also acts as a means of lowering the dependency over the fossil fuel-based generation of power.

The capacity of 'demand/response' helps to match or balance the consumption of electricity with supply. It is further achieved through the automated integration of digital-based metering, for instance appliances and smart meters that recollects and records the data on consumption and output. Such devices help to allow communication (two-way) between all smart grid components with the help of secured I.P. addresses, which further ensures the responses to immediate fluctuations in faults of the system and load-based demands. Also as seen in the previous chapter there are quite a few smart grid technologies which are required to be integrated successfully with the network [31].

Why Smart Grid are better?

The presumption that lies behind the concept of smart grids is to improve the system resiliency, reliability, efficiency, and flexibility of electricity-based distribution system that results in a huge harmful emission reduction[35][41]. These also help in allowing all community areas to become 'prosumers' which involves both consumers and producers of power specified within a community.

The efficiency of the system can improve significantly by the use of smart grids in the form of energy stored [10][38] that can be used for peak-based periods, decreasing the generators size as needed, and thus lowering the use of fuel. The ability to deliver and store energy also helps to enable the sources of renewable energy which are of intermittent form, further lowering the use and cost of fuel. While still in inception, the potential advantages of smart grids to the organisers of the event can be summarised as stated below:

- Reduction of emissions.
- Reduction of fuel costs.
- Renewable energy sources integration.

- Providing real-time and instant energy feedback on the basis of consumption and production.
- Easy fault-based diagnosis on the basis of real-time systems.
- Efficiency-these can help to provide energy based on load demand i.e. storing what is not required on immediate basis for use by other kind of customers within their network set up.

The concept of smart grid is uniformly advantageous for enterprises, universities, retail stores, multinational corporations and hospitals. The whole system of smart grid operates in automated manner to tackle the consumption of electricity at all the places. The architecture of grid is also integrated with control, communication[49][60]. and metering infrastructure[40]. In providing the consumers with present consumption-based information and the information about the prices of energy, the smart grid management services help to curtail the utilization in case of peak-demand times and high-cost[66].

A modernized system of smart grid contains the following abilities:

- It can be repaired on its own.
- It helps to encourage the participation of consumer in the operations of the grid.
- It ensures a premium-quality and consistent supply of power that resists the leakage of power.
- It allows the electricity-market to develop its business.
- It works in an efficient manner.
- It has the capability for energy storage too

A technology of smart grid-based features consists of the following:

- Support of Demand-based response: It provides an automated way for the users to lessen their electricity bills by using electronic low-priority devices as and when the rates are lower.
- Load-based handling: The total grid power load varies over time and is not stable. During heavy loads, a smart grid energy-based system advises the consumers to minimize the consumption of energy on temporary basis.

LITERATURE SURVEY

- Power generation for decentralization: A decentralized form of distributed system of the grid helps to allow a single user for onsite generation of power by engaging any convenient method .

Implementation Features of Smart Grids

The implementation of planned and existing technology of smart grids [11] [47] helps to provide a large range of features for performing the functions that are desired.

a. Load reduction

The load connected in total between energy grids can remarkably change, which means that the global load is not at all stable, slow-varying, average form of energy consumption. Usually, the time responding to a rapid increase in energy-based consumption must be larger and longer than the time of start-up of a bulk generator as most of the unoccupied generators perform on standby dissipative mode. A technology of smart grid, may limit all single devices for reduction of temporary load in order to allow the time for starting up larger generator or regularly in case of resources[27][43]. Along with algorithms based mathematical prediction, it is usually achievable for one to usually point out the standby needed generators used to reach the rate of failure. In conventional system of grid, the rate of failure can be lowered at the cost of increased standby generators. In the technology of smart grid, the load reduction by even small part of clients might improve situations[73].

b. Demand fraction elimination

By using the system of control, the power grid involves varying communicational degrees, as in case of transmission lines, parts of generators, parts of substations and major consumers of energy. Generally, the flow of information is from the user-based sites and the loads, which they control back to the system-based utilities. The utilities of the system mainly attempt to supply or provide the system demand and fail or succeed to the fluctuating degrees as in case of rolling blackout, an uncontrolled form of blackout, and a brownout. The system of demand-response helps to allow the loads and generators to perform real-time interaction. Eradicating the fraction-based demand usually takes place in spikes and removes the addition of reserve generators costs, cuts the problem of wear and tear and it upgrades or extends the equipment's' life, and it allows the users to cut the bills of energy for devices of low priority to use the energy, when the rates are cheap.

c. Power generation distribution

Distribution of generation helps to allow the consumers (individual) to design energy at their location themselves. Such a situation helps to allow the loads of individual type to manage generation directly to their own load, which enables them to become independent from the power grid, due to this, customers can fully avoid the failure of power or energy. The grids of classic form were modelled for the flow of one-way form of electricity. But in case of local form of sub-network, if it generates more energy than its consumption, the flow of reverse type can increase the reliability and safety issues. The technology of smart grid can smoothly manage such kind of situations[104][105].

Department of Energy enlists five fundamentals that run the technology of smart grid:

- Connecting components of electronics, integrated communications in order to get the information and control each and every region on real-time basis, whereas on the other hand, it makes each and every section of smart grid both to talk to and listen.

- Measurement and sensing technologies of the smart grid to provide accurate and faster response of information for each significant section of smart grid such as real time thermal rating, remote monitoring, analysis of electromagnetic signature, demand-side management, and real-time pricing.

- Advanced and enhanced components, for applying the modern research in diagnostics, storage, power electronics and, superconductivity. These generally consists of flexible current transmission (alternating) devices of the system, first and second generation-based superconducting wire, HVDC, high temperature-based superconducting cable, energy generation of distributed form and composite conductors, storing devices, and appliances of "intelligent" type.

- Decision support and the system of information which helps to lower the complexity of grid in order to make both manager and operator to use it more easily and efficiently for the purpose of effective decisions.

LITERATURE SURVEY

- Enhanced methods of control, to monitor the crucial component, enabling fast diagnosis and appropriate precise solutions in case of any event. The three advanced control methods are categorized into: analytical tools, operational application, and distributed intelligent agents.

The new intelligent technologies carry newly built functions of the Smart Grid that makes the grid more competitive than the existing forms of power grid[69][70] .At the same time, this is prone to cyber attacks[6] too hence necessary checks should be in place. Also since this is a promising future technology, the optimization[44] methods for the smart grid paradigm[33] is a good field for research.

2.3 Parameter Estimation

In 1970, pioneering work conducted by F.C. Schweppe consists of the work done on static estimation (SE). It plays a significant function in supervised planning and control of electric power grids. It helps to screen the grid state and allows EMS i.e. energy management systems to accomplish numerous significant control and planning tasks like establishment of nearby network models on the basis of real-time for proper functioning of the grid, bad data analysis/detection, and optimization of power flows. One more SE-based utility example involves SE-based assessment of security/reliability that is basically deployed for analyzing various contingencies of the system and it determines the desired preventive actions in power systems research area. In regard of current development of smart grid, a good research on SE [25] [50][55] is generally required for meeting various challenges of smart grid-based functionalities present in the system. In this work, a brief summary of few technologies of SE which have matured over the past decades has been provided, this basically helps to examine the opportunities and challenges presented to the evolution of the grid into a smart form, within its relevant framework.

Evolution of State Estimation

The power system state is generally explained with the help of phase angles and voltage magnitudes at each and every bus of the system. Such an information with topology-based knowledge and grid-based parameters, can be used for characterization of the entire grid system. SCADA (Supervisory Control and Data Acquisition Systems) represents a set of tools, which are used for monitoring, controlling, and optimizing the power system performance. SE presents a crucial component of the system. SCADA attains the

LITERATURE SURVEY

measurement from various devices such as RTUs i.e. remote terminal units and in more recently, PDCs i.e. phasor data concentrators. The state estimator (SE) helps to calculate the state of the system and provides the basic desired information to SCADA, which finally takes various actions by forwarding control signals to the switchgear[4]. The conventional SE built into energy management system (EMS) generally involves four major processes. The processor-based topology generally tracks the topology of network and it upholds the database on real-time network-based model. The analysis on the basis of observability represents a process, which basically ensures the set of measurement, which is tolerable to perform the state estimation. In the next step, the processor based on bad-data helps to identify the gross type of errors that take place in measurement and it removes bad type of measurements. The state estimator (SE) begins its operation on a set of good measurements once in order to evaluate the system state. Lastly, the processing of bad-data determines any kind of gross errors in the measurement set and removes the bad measurements. Depending upon the evolution and timing of the state estimates, the schemes of SE can be categorized into two of the basic unique paradigms. The first one is the static SE i.e. SSE and the other one is FASE, namely, forecasting-aided SE.

Static State Estimation

From past four decades, there occurs a large amount of SE-based research and the SE has focused over the technology of SSE[102] which relies over the fact that the conventional techniques of monitoring, like the one that was realized in SCADA system only takes place using non-synchronized type of measurements after every second i.e. two to four seconds. Later, in order to lower the complexity of computation needed in implementation of SE process, the estimates of the system are generally enhanced every few minutes only once. Hence, the SSE usefulness is meant to administer the monitoring on real-time basis and it is quite restricted in practice.

2.4 Phasor Measurement Units

The concept of Phasor measurement units (PMUs) are considered a valuable component of wide area measurement systems utilized for safe grid operation. PS complexity is rising on daily basis because of load-based growth, addition of generation at several places, which involves the process of grid-based integration of electricity market mechanism (EMM) and renewable energy sources (RES). The corridors of power transmission are needed to run to its maximum limits in order to coordinate with the system's fluctuating load and power

LITERATURE SURVEY

injection-based statistics. This endangers the system security and online monitoring. The operational strategy of the system helps to evaluate the major challenges in wide area power network. In order to maintain the operational reliability as well as the stability of the grid, synchro phasors are generally adopted with various promising results for obtaining better performance of the system. It has been observed that the knowledge achieved by the phasor measurement units (PMUs), helps the operators at large for situational visualization and awareness in wide structure of grid[24]. The impressive and effective process of utilizing the PMU information in order to enhance the efficiency of operation has been illustrated by the system utilities. Estimation and Monitoring of several states at the system bus represent a key component of trendy EMS i.e. energy management systems[57].

The power system state is normally categorized as the assortment of positively sequenced currents and voltages in the system network. Control of wide grid and real-time monitoring with multiple load connectivity and power injection devices in a diverse manner can be completely gained with precise deployment of PMUs. Protection, control of the grid, and monitoring can be effectively done with wide area management system. The PMU acquired high resolution data is adequate to acknowledge the dynamically built system response and it also helps to find out the real sequence of the event. The traditional form of SCADA system transmits the data information from the Remote Terminal Units that are placed across distinct generating stations and substations. The quantities like voltage, reactive and real power, circuit breaker, and switch location are either reported by periodic scan or by exception periodically.

2.4.1 Power System Monitoring with PMU

The implementation of PMU has set forth a shift from state estimation (SE) to the determination of state. In India, the experience of synchro-phasor is enhancing and of promising nature and it has prompted a magnificent comprehension of grid framework. Precise time stamped information has empowered administrators to make a notable contribution to the grid. The utilities are probably going to be profited with cutting edge static and dynamic contingency-based investigation. PMUs represents the PS units that gives synchronized estimation of constant phasors of current and voltages utilizing GPS[46][48][61][67]. The time stepped signal encourages the control, coordination, and monitoring of huge system by method of signal examination. Fig. 2.4 below represents PMUS-based block diagrams. The technology of Synchro-phasor is generally comprised of

LITERATURE SURVEY

PDC, PMUs, communication network, historian, offline toolboxes, and visualization on the basis of real time. From PMUs, the timely synchronized data is collected by PDC. It also analyses the information and the quality of time sync and feeds such data to the historian. The main function of the Historian is to archive the data for few years depending upon the capacity of storage. Such type of data can also be utilized for forensic analysis and post disturbance analysis. The communicational network generally consists of infrastructure of wide band communication (high-speed) from the substation to controlling centres. The applications of technology in power systems are based on synchro-phasor and can be smoothly grouped as advanced protection of the network.

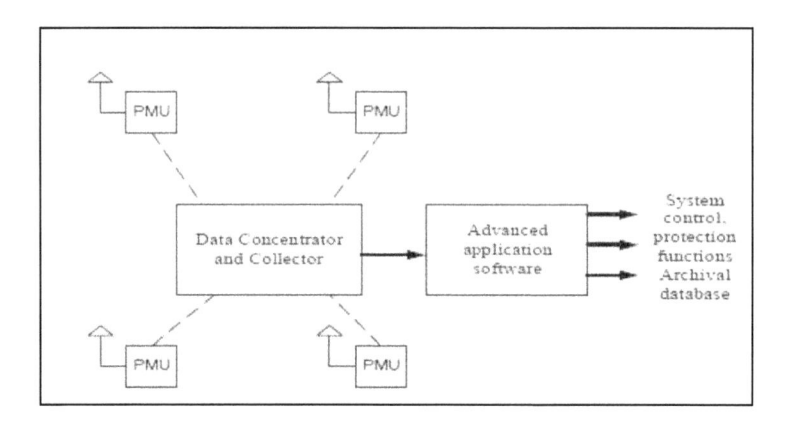

Figure 2.4 Utilization of PMU in Power System

2.4.2 Analytics and Control of PMU Signals

The PMUs are used for signal based detection . Although, the amount of received data is very large and requires a suitable analysis in order to conform to a controlled decision. The PMU-based signals are voltage, current, phase angle, frequency and rate of change of frequency. On basis of real time controlling application, the suitable or accurate signal has to be pointed out for fast regulation of the system[29]. Therefore, the analytics of data may be needed for identification of signal and then further for processing in case of controlled decisions. On the other hand, an approach based on algorithm needs to be unfolded for the process of signal detection and data mining for understanding the system health signature analysis for assistance to the operator. This enhances the research area for the analysis of the signal and the relating applications with the use of PMUs[39]. A survey and study is done over the PMU-based application in Power Systems. The WAMS applications can generally be grouped as advanced controlling scheme, real time monitoring in power systems and advanced protection of network.

2.4.3 Optimized PMUs Placement in Power System: There are PMU-Based placement algorithms for the state estimation (SE) under single branch outage and measurement loss. A heuristic technique is also presented for rearranging the position of measurement for minimization of the PMUs number. Several methods for optimal placement of PMUs for state estimation of the power system are presented [7][8]. They generalize on the basis of measurement uncertainty while observing a minimized number along with PMU-Based optimized location for the purpose of state estimation. An approach for optimal placement of PMUs requires a complete observable procedure in its normal state of operation as well as the observation for an individual branch outage[16][17][76]. The algorithm based on Binary search is also used for having minimized PMUs (number-based) for the process of observability, it takes a huge time for execution. Bei Gou (2008) has stated [59] a generalized form of integer-based linear programming method for an optimized placement of phase measurement units (PMUs) considering redundancy of PMUs location, incomplete and full observability process. Such type of model leaves all impacts of zero injection buses. But it has been extended/enhanced in order to observe the effect of zero injection along with the process of redundancy measurement. A specific method for finding out the minimized PMUs number and their optimized positions for enabling the whole network system to become observable. This kind of methodology grants the loss of multiple or single PMUs enabling the whole network of the system to be observable. The analysts in [23] has provided an approach based on the immunity-based genetic algorithm (GA) for optimized PMU placement for the system to become completely observable. This methodology represents optimized placement of PMUs without any contingency-based measurement.The research also show that if placed optimally the PMU can help in estimating unmeasured data also[18].Hence, immense research has been done optimum PMU placement using different algorithms and approaches.[120] to [130]

2.4.4 PMU for State Estimation: The state estimation of power system utilizing PMUs has been inspected on critical basis. The inaccurate measurement and the model of the system generally represents the system that gives rise to anxiety or uncertainty. The function of state estimation represents a basic tool for real-time data measurement received from the controlling substations. AliAbur et al (2005) investigated [62] the problem of SE in multi-area systems. It indicates a central type of entity coordinator which presents a possible solution of SE-based area using data from measurement of synchronized phasor from each of the bus and the raw-type measurements from boundary area. M. R. Irving et al 2005

LITERATURE SURVEY

overviewed a comparative analysis among two of the methods for the process of PS state estimation with intervals of uncertainty [64]. The formations of linear type and nonlinear constraints are proposed for estimation of the possible tightest bounds of states considered. This kind of uncertainty is constructed via lower and upper bounds on the basis of measurement of errors. A state estimation (distributed) is viewed for power system on large scale basis, the overall system distributed algorithm generally disintegrates into small newly built sub-systems and each of the sub-system derives their own estimation of state and provides a local from of state estimation output. Luigi Vanfretti et al 2013 analysed a [48] method for phase measurement unit on the basis of state estimation terminology that functions to perform the correction of phasor angle bias along with redundancy measurement. The method for possible solution of state estimation on the basis of PMUs are provided (given). Static state estimation (SSE) algorithm is of iterative nature and begins its operation as a flat start, which uses bulk computational analysis and it is not able to be executed in intervals of small type.

Also sampling rates of measurement from SCADA systems is different from that of PMU hence hybrid approaches have been researched to handle this time skewness [22].

2.4.5 PMU Application for Assessment of Voltage Stability: In a recent analysis, the monitoring process of system stability and voltage using the technology of synchro-phasor has obtained a wide range of acceptability because of time-stamped system features [5]. The algorithms based on PMUs are used to pin-point the proximity of voltage collapse. The process of ranging the methods from easy static methods to hard dynamic approaches have been studied and described voltage stability assessments (VSA). Amethi, R., et al 2013 presented a network of PMUs for wide areas for control and monitoring voltage stability [46]. They have discussed an idea for the detection of voltage instability and corresponding control for voltage dependent loads. Ruisheng Diao et al 2009 provided [74] an approach for assessment of online security of voltage security using decision tree and PMUs, which gets trained (offline) by the analysis of voltage security with its set of representatives and 24hours conditions of forecasting. The attributive values resulted from PMUs on the basis of real time gets compared with threshold offline that is usually determined for assessing the system security [19].

2.4.6 Fault detection/location with PMUs signals: Pandey, R. K. et al 2014 have provided an adaptive approach of PMU for the purpose of detecting and locating the faults on an

UHV/EHV line of transmission accurately [23]. Such an algorithm is fast, suitable, and recursive, which may be further applied in computer relaying field. C W Liu et al 2002 overviewed [88] a newly built algorithm on the basis of fault location using PMUs for lines of series compensated form. Such a method uses two-step based algorithms specifically, a step that is pre-located and the step corrected, for evaluating fault location along with drop of voltage. This is easily applicable to any of the series-based FACTS. Chi-Shan Yu 2002 has provided an adaptive form of mimic-based phasor estimator so as to do away with the dc oscillations decaying between voltage and current [71]. A mimic phasor estimator on the basis of DFT is usually developed with an adaptive form of scheme in order to gain the decaying time constant.

2.4.7 PMU Application for Control and Monitoring: The measured utility of the signal occurs in centralized control manner from the locations of remote type with wide area management systems (WAMS). A predictor-based on design of H-∞ control is generally used for modelling the delay of signal transmission, on the basis of SVC-WAMS damping controller. Here, constant delay of signal transmission is usually taken for all the channels of communication that may get changed in real practice. A monitoring method of PMU based power system stability is provided in terms of Fourier spectrum for estimation of the eigenvalues that gets changed with the global generation and the area-based power flow is generally discussed. The monitoring done on the basis of inter-area-based mode of oscillation, limits of stability and Available Transfer Capability (ATC) are evaluated. The modelling process of WADC, which is popularly known as wide area damping controller for the case of inter area-based oscillations is highlighted by Yang Zhang et al [73] [74] [79]. Here a procedural formal design/model of wide area damping control (WADC) systems is modelled by jointly combining the LMI i.e. linear matrix inequality on the basis of robust control model and a stabilized signal selection. However, the use of continuous, linear approaches helps in designing of the controller representing a very strong and powerful methodology, it also needs appreciable process of tuning, which is generally followed by the criteria of testing.

Expected Benefits of PMUs

- Helps to provide wide-area-based circumstantial awareness for the operators of the system.

- Aids to determine the possible margins of the system.

- Helps to determine the points of stress with respect to the system of transmission by the process of monitoring the quantities of phasor.

- Aids and detects to restore the section that is islanded from the grid after a major outage or storm causing system disturbance.

- Provides analysis of post-disturbance capability.

- Enables PMU data-based visualization for the system operations to get incorporated into EMS.

- Improves the accuracy and state estimation (SE) of various applications of EMS as the direct form of data is more authentic and it overcomes the delays related to the process of modelling.

There have been various studies with respect to optimizing PMU usage in Power Systems using state estimation techniques, reduction of the nonlinear component in the parameter estimations, the reduction in the number of PMU's used and their optimum placement.

Helder RO Rocha, et al [111] presented a new approach for designing the WAMS i.e. Wide Area Measurement Systems. An algorithm based on topological analysis in the basis of variable neighborhood search heuristic was tested and proposed in various type of networks, that includes IEEE testing networks and the Brazilian 5804-bus transmission system. The analysis of the results has shown the effectiveness, scalability, and flexibility of the proposed framework on comparing it with presented literature-based research.

Morteza, Sarailoo et al [112] was to accurately infer the isolated PMUs-based synchro phasors in order to hide the spoofing effects on the basis of real-time functions in respect to the network of PMU. As presented in the framework, a solution of deterministic form usually requires more redundant form of PMUs and lots of interconnections in the environment based on the circuits of physical transmission. Therefore, the cost (defensive) rapidly develops with increased amount of attacks based on aggressive spoofing. In order to address this kind of problem, various directions have been considered.

Zeina Al Rammal, et al [113] discussed the optimal phasor measurement placement (OPP) for reverse-based detection of power flow. A comprehensive review of literature and a comparison among a large range of already existing algorithms of optimization was done.

LITERATURE SURVEY

Further, genetic algorithm (GA) was selected for solving such kind of problem. MATLAB based Global Optimization Tool was used for testing the algorithm that was proposed on IEEE-39 and IEEE-14 node-based test feeders.

Xingzheng Zhu, et al [116] proposed the problem based on OCLP i.e. optimal PMU-communication link placement which investigates the placement of communication links (CLs) and PMUs for full power system observability. Besides the location of CLs and PMUs, in order to ensure there timely and liable PMU data transmission, the capacity of communication required on each and every CL was also captured by the problem of OPLP. The researchers have carried out the numerical-based studies on the IEEE 300-bus, 118-bus, 57-bus and 30-bus systems for the model proposed. The results have shown that on comparing with conventional optimized model of PMU placement, OCLP can help in reduction of the significant total cost of installation.

Saroj Kumari, et al [121] have discussed a technique for placement of PMUs so as to be able to observe the complete power network with minimum number of PMUs, since the installation cost is significant. They have presented a BPSO which is binary particle swarm optimization and is implemented over Puducherry 17 and a standardized IEEE bus system. The said technique can be used for any of the power systems to make the system completely observable for different requirements of the power system. The results were compared with a few existing methodologies and it was discovered that the Adaptive GILP, SA, GA BSA and the proposed method of BPSO were more efficient

V.Vijaya Rama Raju, et al [125] proposed a methodology on OPP i.e. optimal PMU placement problem. It was articulated as a BILP (binary integer linear programming) using BAA (Balas additive algorithm). The installation of PMU was further decided using binary decision-based variables i.e. 0, 1 for full observability of network while minimizing the number of installations of PMU. It also considered the zero injection buses for reducing the PMUs number . The connectivity of power system matrix was mainly presented in a binary form and simple methods of heuristics were used for solving this issue. The problem related to optimal placement of PMU possess multiple kind of solutions having equal costs. For the process of ranking such multiple kind of solutions, the method of redundancy measurement was used. The algorithm proposed has been tested on IEEE bus systems i.e. 30-bus, 24-bus, 14-bus, 9-bus systems.

LITERATURE SURVEY

Jyoti Paudel, et al [127] discusses a technique for strategic locations to install additional Phasor Measurement Units (PMUs) while determining the system resiliency of the existing PMU based measurements. An virtual attack agent was modelled using an optimized framework. This virtual attacking agent could minimize the power system based observability using a coordinated form of attack which was then used for a subset of critically designed PMUs The framework was applied on 118, 57, 30, and 14 bus-based test systems.

George N. Korres, et al [128] describes the technique of the various methods of optimization which were used for the problem of OPP hence analysing and categorizing the on going research in this field. The techniques were divided into two types: mathematical algorithms and heuristic algorithms The commonly used methods for the optimal solution of the OPP issue are the IP and SP algorithm. The bus systems used were the IEEE (14 to 300) bus systems. The report was presented for the Polish power system which is a fairly wide spread system containing 3514 lines and 2746 buses. MATLAB was used for implementing most of OPP methodologies.

M. Hurtgen, et al [98] proposed to minimize the PMU configuration while being able to observe the network completely The method suggested a PMU distribution to observe the network. The ILS which is iterated local search is metaheuristic and is used for reducing the PMU size configuration required for observing the network of system. This algorithm was tested on IEEE testing networks with 118, 57, and 14, nodes and it was further compared to highlight results.

Saikat Chakrabarti, et al [96] described a method for synchronized measurements for the complete power system observability. PMU placements utilize the time-synchronized measurements of current and voltage phasors. The approach was called the integer quadratic programming and was used for reducing the required total PMUs .This further maximized the redundancy-based measurement at the power system buses

P.S. SreenivasaReddy, et al [89] presented an extensive study for optimized placement of phasor measurement unit (PMU) for the increasing complexities in a power system. This advocated the use of a distinct type of algorithms for optimum placement of PMU .

B. Mohammadi-Ivatloo, et al [80] presented an algorithm based on optimized PMU-based placement for the observability in power systems and it also enhanced the performance of

LITERATURE SURVEY

the secondary voltage control scheme. Here the branch and bound method of optimization was utilized for solving the problem of OPP which was applicable to the problems with Boolean and integers variables.

Chunhua Peng, et al [75] optimized the placement of PMU for full power network observation and the minimized number of PMUs. It usually provides a speedy and a general method of analysing the topology of power network observation on the basis of properties of phase measurement unit and structural information as per the topology of the power network, and resolution of the object function by improved version of binary PSO algorithm that was further combined with a mechanism on the basis of information based on immune system information.

Madhavi Kavaiya, et al [54] presented a methodology based on integer based linear programming for optimal PMU placement in a known network for achieving the full network observability. Firstly, a complete conventional observability of the given network was mainly designed and then the bus constraints based on zero were added in previously designed formulation. The results obtained from modified and conventional formulation were compared further. However, minimized problem of PMU placement may contain multiple network solutions, so in order to decide the best solution, two of the indices were proposed, SORI and BOI, where SORI is System Observability Redundancy Index and BOI is Bus Observability Index. Results over IEEE 14, 9 bus were presented.

T. L. Baldwin, et al [34] concerned the extension spanning tree' concept in respect of measurement of spanning subgraph with a pseudo or an actual type of measurement mainly assigned to each of the tree branches. For acceleration of the adopted procedure, an initial placement of PMU was provided that builds a subgraph on the basis of spanning measurement in accordance to a method of depth-first search. From the results based on simulation on various testing systems, it usually appeared that only one fourth to one third of the system was required to be taken care of for making the system to be observable.

In the table below a comprehensive comparison of the various approaches studied has been presented

Table 2.2 Comparison of different approaches

Authors (years)	Topologies	Technologies	Algorithm
Helder RO Rocha, et al(2019)	Wide Area Measurement Systems	Variable Neighbourhood Search	Metaheuristic
Zeina Al Rammal, et al(2018)	Reverse-based detection of power flow	NA	Optimization was done for the purpose of research work analysis.
Xingzheng Zhu, et al(2018)	Communicational links	PMUs relocation and distinct paths of transmission	BILP (binary integer linear programming) using BAA (Balas additive algorithm)
V.Vijaya Rama Raju, et al(2016)	NA	PMU based measurement system resiliency	Analyse the critical form of PMUs
Jyoti Paudel, et al(2015)	Complete power system observability	NA	Programming was utilized for reducing the total number of PMUs.
Saikat Chakrabarti, et al(2009)	NA	NA	Phasor measurement unit for enhanced development of the power system.
P.S. SreenivasaReddy, et al(2010)	Secondary voltage control performance	NA	Optimal placement problem is designed to reduce the number of PMUs installed

LITERATURE SURVEY

			while ensuring full network observability and also tracking pilot buses of the system for encouraging the secondary voltage control performance.
B. Mohammadi-Ivatloo, et al(2008)	Placement of PMU for full power network observation and the minimized number of PMUs	NA	Binary PSO algorithm that was further combined with a mechanism on the basis of information based on immune system information
Chunhua Peng, et al(2008)	NA	Conventional observability	SORI and BOI, where SORI is System Observability Redundancy Index and BOI is Bus Observability Index. Results over IEEE 14, 9 bus were presented
Madhavi Kavaiya, et al(2008)	Observability was implemented on the basis of topological analysis	NA	Discrete particle swarm algorithm binary version was mainly used as a tool for optimization so as to find the minimum number of PMUs for achieving the observability of the entire system

Also in Table 2.3, a comparative analysis between different methods of OPP has been done on the basis of various characteristics.

LITERATURE SURVEY

Table 2.3 Comparative Analysis Between Different Methods of OPP

No.	Characteristics	Algorithm based on Programming		Heuristic Methods		
		Integer Linear Programming	Integer Quadratic Programming	Matrix Reduction	Genetic Algorithm	Particle Swarm Optimization
1	Computational Efficiency	High	Moderate	Moderate	Moderate	High
2	System Observability	High	High	High	High	High
3	Computational Time	High	Moderate	Less	Moderate	Moderate
4	Complexity of system	Low	High	Low	Moderate	High
5	PMU needs	Moderate	High	Low	Low	High

2.5 Kalman Filter

The data of PMU usually captures the response of the power system created by distinct types of events based on power system. Hence, on using PMU's, the system becomes completely observable PMU's measure the data in real time. For a smart grid to be a success, we need to achieve dynamic state estimation too[20][21][28][55]. Numerous methods of Kalman Filter have been used and applied for extraction of the components of steady state in context to PMU based measurements [97][101]. The applications of such methods make them capable to reduce noise, compensate for hidden or missing data and filter outliers from PMU based input signals. This is another avenue of research to be used as an advantage for the implementation of a smart grid. Optimal filters have been designed using Kalman[72][77][80], Extended Kalman[28][36][58][84] and Unscented Kalman filter[42][45][51][52][53][56] algorithms to make an accurate estimation of the power system parameters, neglecting the random or measurement noise. Initially, using synchronized phasor measurements which are provided by PMU, a linear estimator is formulated and subsequently work is to be done to remove the nonlinearity.

LITERATURE SURVEY

Jinghe Zhang, et al [114] in his work combined the dynamic estimation method of a Kalman filter with PMU so as to improve the quality of data used in smart grid. He further compared the noise variances of a classic Kalman filter and an adaptive Kalman filter

Xin Wang, et al [115] discussed that for real time state estimation to be effective, it is important for the synchronization of the smart grid with the rising demand and renewable energy sources. Hence to check the unwanted data for PMUs, they presented a novel nonlinear Kalman filter based estimation framework in which rather than using direct analysis of abc coordinate frame instead a symmetrical component transformation was used to separate the positive, negative and zero sequence networks.

Junbo Zhao, et al [117] worked on a robust iterated extended Kalman filter which talks about a maximum likelihood approach (termed GM-IEKF) to estimate the state dynamics of the power

System when subjected to disturbances.

Hadis Karimipour [118] had proposed a two level lateral dynamic state estimator using the extended Kalman filter. This used both SCADA and PMU measurements. For the bus measurements which did not have PMU installations, predictions were based on previous data. Subsequently they were then compared with a multithread CPU based code.

A. Sharma, et al [2] presented a theory for tracking power system state estimation using a cubature Kalman filter. This worked on the synchronized phasor measurements from the various phasor measurement units and subsequently, a forecasting technique was used to forecast the states of the period when PMU data was missing due to temporary snapping of the communication link.

E. Ghaheramani and I. Kamwa [58] worked on the extended Kalman filter technique for dynamic state estimation while making use of PMU quantities. They proposed the extended Kalman filter with unknown inputs referred to as EKF-UI. Simulation results depicted the efficiency of the method under faulty conditions and has the potential to be used where there was no information of the input signals of the system.

Jinge Zhang, et al [119] in his work presented a two stage Kalman filter used to estimate the static state of voltage magnitude and phase angles. Kalman filters depict a reliable performance when the system noise be characterized by some stastical properties. But rarely one is completely able to define the process and measurement noise models. Therefore he used an adaptive Kalman filter with inflatable noise variances (AKF with In No Va) which decreased the influence of wrong system modelling.

Kalman filter is a linear quadratic estimation method, hence in most cases cannot handle the non linear data. For this extensions of Kalman filter like EKF [15][28], UKF and adaptive filters are used.

2.6 Optimization Techniques

This part provides a brief summary of the optimization techniques whose usage depends over the space solution and optimization complexity issues.

There are various optimization methodologies, conventional and heuristic methods. Some of them are discussed here.

Conventional Methods

(i) IQP: In the analysis of Integer Quadratic Programming (IQP), the problem of Optimal Phasor Measurement Placement (OPP) is usually described by using the matrix-based connectivity that denotes topology of the network. The optimization of quadratic objective function that considers the linearized constrained along with consideration of integer-based variable value. This type of process will help in reduction of PMUs number by making a system full observable under normal operating and outage conditions.

(ii) ILP: Integer Linear Programming techniques, usually called as Binary integer programming, considers both the power calculation and injection of the system additionally with failure error calculation of PMU with the help of state estimation (SE). The process of ILP formulation is usually based on the values of eigenvectors usually obtained from the adjacent matrix of spanning tree. In order to curtail the rate of installation, after the process of decomposition, using the ILP theory, the PMUs are optimally located in the sub networks of the system.

(iii) Greedy Algorithm: An algorithm based on combinable optimization that considers an appropriate form of local, or immediate, possible solution while searching the answer, is known as GA i.e. greedy algorithm. An algorithm based on matrix reduction and a virtual data pre-processing elimination approach have been introduced for decreasing the computing effort and placement model size for determining the set of optimized placements.

Heuristic Methods

As, the installation of PMUs is not of cost-effective nature, hence at each of the bus to control and analyze the area of fault, it is necessary to implement distinct techniques of placement that will help in providing a complete system observability. Presently, there are various types of heuristic forms. We will discuss various types of heuristic methods as stated below:

(i) Particle Swarm Optimization: PSO represents a technique of stochastic optimization based on population. With the help of such kind of methodology, it is really feasible for mapping the design criteria of configuration along with modeling the data-based loss. Benefit: Offers solution in multiple forms against the losses of data at the notable PMUs.

(ii) Genetic Algorithm: GA provides a full observability of power system with minimized units of PMU along with geological distribution. It usually takes the relation between the units of PMU along with current phasors technology. The specific advantage of this kind of methodology is to provide best Pareto optimized solution instead of a single solution.

(iii) Tabu Search: TS represents a local search algorithm based on meta heuristic technique. This methodology is applied for solving the problems of combination-based optimization by guiding and tracking the search. The process of TS usually comprised of incidence matrix manipulating the integer numbers and TS algorithm. Benefit: More proper results with quick computational time.

(iv) Artificial Neural Network: ANN is capable of providing an approximated functional relation between system parameters and voltage stability indices. Benefit: Helps in providing multiple kinds of solutions on the basis of computing models.

(v) Simulated Annealing : SA presents a method based on concept of genetic probabilistic meta heuristic methodology. The main motive is to approximately search a solution (reasonable) regardless of its best possible solution in a fixed interval of time. A method based on the analysis of practical sensitivity was considered for calculating the sensitivity parameter of each and every bus. The methods based on incidence matrix are being used for

LITERATURE SURVEY

placement of initial observability of PMUs. Benefit: This type of method helps in providing power system full observability system in association with valuable form of dynamic measurement of data at the same time from the power systems.

(vi) Mutual Information: The concept of reducing the number of PMUs with the help of an information-based theory, popularly known as Mutual Information Benefit: The uncertainty of the system can be easily modeled.

(vii) Ant Colony Optimization: ACO is a technique of classical probabilistic form that mainly uses graphs reducing the problems of computational complexity. The main advantage of such type of systems is to reduce the complexity of the system.

(viii) Iterated Local Search: It represents the merger of both the ILS and PPA i.e. page rank placement algorithm. It benefits an ease of implementing as well as understanding the conceptual framework.

(ix) Exhaustive Search: ES is a well-known method to be called as exhaustive as this type of methodology provides an assurance in providing all the reachable states before it gets declined with unit-based decline. Benefit: System failure can be easily detected.

(x) Reduction of Matrix :This type of method helps in minimizing the number of PMUs with the help of a method by using matrix reduction technique. And it further solves the problem of OPP on the basis of mathematical approaches such as pre-processing of virtual data elimination, Lagrangian relaxation or algorithm based on matrix reduction. Benefit: The computing time will decrease making it easier to be applied in wide scale energy systems for achieving full system observability

Exact Methods

The exact methods are popularly known as complete methods, these are generally capable of searching an optimized solution to a big issue. They usually range from various techniques that are considered to be very useful for solving a large category of optimization-based issue like Branch and Bound (B&B) [86] [90] to the algorithms that are problem dependent like Dijkstra's algorithm [82] [83]. The exact general exact method of optimization can be categorized on the basis of the subsequent paradigms of Heuristic (A*).

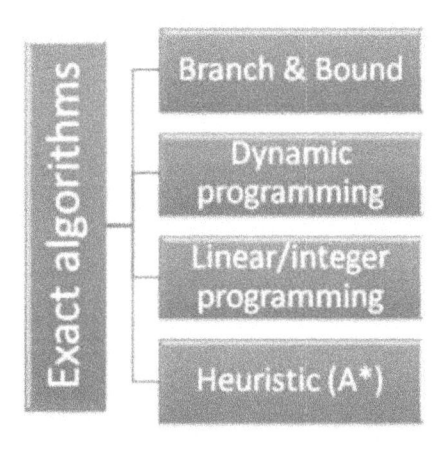

Figure 2.5 Exact methods of classification

(i) Branch and Bound: The Branch and Bound method was proposed and planned by Doig and Land in 1960 [Doig and Land, 1960]. A B&B algorithm [82] involves enumeration of candidate-based solutions in a systematic manner with the help of a search (rooted) tree. This proposed algorithm helps to explore the branches of a tree that represents the sub-sets of solution-based set. Whereas the proposed algorithm functions to explore only the branches, it further checks them against lower and upper bounds estimated on the optimized form of solution. Finally, the branches in the process gets discarded if these do not contribute to have better form of solution rather than the best form[87][99].

(ii) Dynamic Programming: Dynamic programming [Bertsekas, D. P. (2005)] is used for processing a complex problem by reducing it into a sequence of smaller, and therefore easier, sub-problems [63]. Then the solution of the larger problem is arrived at by debugging the individual smaller problems.

(iii) Linear/Integer Programming: The concept of linear programming is generally used for finding a solution to various problems of optimization that are presented by an objective linearized function and linearized constraints of inequality. The space solution represents a convex type of polygon demarcated by various constraints (planes). The concept of Integer programming is generally based on the method of linearized programming methods. Unlike the concept of linear programming, the programming of integer form accepts that there are few or all the type of variables that are confined to be of integer form rather than the real once. The problems related to integer programming are mainly solved with the help of [70] branch and cut algorithms [Mitchell, 2002]. Such kind of methods helps to jointly combine the B&B algorithm with the methods of cutting plane. The methods of cutting plane are used for adding various constraints to initialized issues in order to improve the problem of relaxation to a more closely formed problem to approximate the programming based on integer concept.

(iv) A* Heuristic Method: A* (Hart et al., 1968) represents the method of Dijkstra's algorithm extension. It consists of a graph i.e. goal-directed on the basis of traversal strategy being able to find the optimal solutions-based least-cost path from a known node source to a node of target form. It involves order of distinct possible paths and firstly it explores the most suspicious one. The estimation of the process is done by the algorithm that helps to estimate the quality of each and every path that uses a knowledge-based heuristic form of cost function. The knowledge section represents the going cost from source-node to current (present) node, and the heuristic represents the cost of estimation going from current to the target (destination) node. The heuristic method has to accomplish desired set in order to assure optimized solution. A* [78] [81] is considered to be the encouragement of most of the variants like IDA* [Korf, 1985], D* [Stentz, 1994] etc.

Inexact Methods

The concept of inexact methods involves the type of algorithms which return a reliable solution on the basis of finite time along with absence of solution-based accuracy that means it is not all known if in case the solution returned represents an optimal, bad or good type of solution [85] [86]. These algorithms are generally known to be approximate algorithms. Figure 2.6 helps to provide the method of classification in accordance to the used paradigms by the inexact type of methods in order to perform the operation of search. On this basis it can be identified as per the following methods:

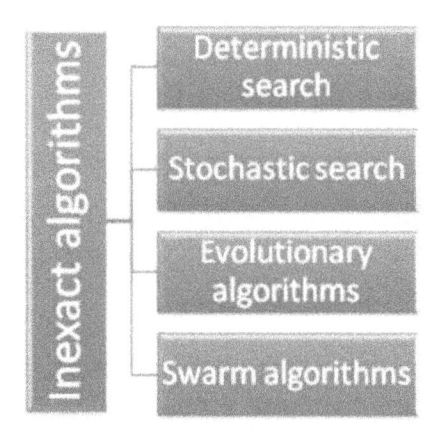

Figure 2.6 Inexact methods of classification

LITERATURE SURVEY

(i) Evolutionary Algorithms: Evolutionary algorithms employ principles of biological evolution to solve optimization problems [79]. They start with a population of solutions which evolve, improving the quality of the solutions, throughout a finite number of iterations. Most popular examples [65] [79] are Genetic Algorithm (GA) [Torrent-Fontbona, 2012, Holland, 1975], Memetic Algorithms [Moscato et al., 2004] and Immune Algorithms [de Castro, 2002, Cai and Gong, 2004].

(ii) Swarm Algorithms: Such type of algorithms are usually encouraged by collective form of intelligence. These are mainly presented as the systems that are of decentralized form having simple homogeneous-based systems that interacts in a local manner with environments. Even though the absenteeism of the structure based on the centralized agent-control method, local type of association among them generally results in a behavior of global type in regard to the operation of the entire system . Some instances are Particle Swarm Optimization (PSO), Ant Colony Optimization (ACO). Despite the process of classification given by Figure 2.6 above, most of the algorithms uses more than one of the mentioned paradigms in the given figure. For instance, stochastic form of search is generally used by most of the evolutionary or swarm algorithms like GA or PSO. Figure 2.6 must not be taken as algorithm of strict classification, but as a paradigm representation generally used by the inexact methods. Evolutionary, Swarm, and Stochastic algorithms are also called as the meta-heuristics methods. The concept of meta-heuristics is generally used to refer to the algorithms in general (making few of the presumptions based on the problem that is to be solved) that engage randomness degree in order to search the near-optimal or optimal solutions to the complex issues. The approaches presented in this dissertation is generally based on meta-heuristics technique. Therefore, such a section usually provides a deep study. These can also be categorized based on the following strategies:

Non-Nature Inspired vs. Nature-Inspired: Based on various origins, the concept of meta-heuristics is generally considered as the algorithm i.e. nature-inspired such as SA and GA, or the algorithms that are non-nature-inspired such as iterated local search or tabu search. However, this presents an easy form of metaheuristics, there are many of the algorithms that fits both the classifications, as they based on the inspired form of natured and non-natured concepts. But this inspired algorithm do not present any particular merit or demerit while solving the problem of optimization. Thus, [68] such a method of classification is considered to be useless to handle a specific problem [Roli and Blum, 2003].

LITERATURE SURVEY

Single Point Search vs. Population-Based: Based on the solution number, the concept of meta-heuristics is basically categorized as algorithm based on the population search such as GA and PSO, or algorithms on the basis of individual point search, like SA. The algorithms that work on individual solutions generally explains a solution space trajectory, the algorithms that are considered to be population-based works with solution sets interacting among themselves. With the algorithm that is based on population search generally provides a flexible methodology to explore the searching space. Although, performance (final) is strongly dependent on manipulated form of population search.

Dynamic vs. Static Objective Function: Most of the meta-heuristic's methods let the objective-based function to be given in the form of problem-based invariable representation along the process of search. Though, the other form of meta-heuristics such as search of local form, mainly modifies the function of objective during the searching methods to get escaped from the local points of optima [Birattari et al., 2001, Roli and Blum, 2003].

2.7 Inferences out of the Literature Review

The study and analysis based on the literature survey has been done. Some of the inferences from the literature surveys are discussed below.

- For Smart Grid to be successful, real time dynamic state estimation of parameters is important.
- Phasor measurement unit helps in state estimation.
- The use of PMUs makes the system observable but as the number of PMU's increase, they inflate the noise dynamics of the system too and also add to the cost. Hence it is important
  to use an optimum number of PMU and place them appropriately to make the entire system observable. Already lot of research is going on in this field.
- To reduce the non-linearities in the system parameters due to the enhanced noise dynamics
  caused due to the use of PMU, the Linear Quadratic Estimation method of a Kalman filter when combined with PMU data can help in dynamic state estimation.
- Besides, the process of sampling procedure in PMU does not fix the size of sampling so random sampling reduces the accuracy of data because noise increases

LITERATURE SURVEY

- Based on the analysis of literature, all the methods of PMU parameter mainly satisfy the adopted criterion. But the speed based on the convergence of filter is not fast as it increases the overhead and reduces the accuracy and increases MSE.
- On comparing various cases in terms of analysis based on with and without communicational links (CLs), the experts have mainly found the presence of already existing forms of CLs that may result in the PMUs parameters and distinct paths of transmission for PMU-based data. Finally, the experts have found a very small change in CLs or PMUs prices which influences the optimal results in a significant way that proves the optimal PMU-communication link importance in terms of saving cost.
- The problem decision of number of PMU that was usually considered as a problem of a zero-one problem of linear programming (LP) was ended with distinct multiple solutions, not with a specific kind of solution.

2.8 Problem Formulation

The main issue in power grid systems is the problem of load flow. Smart Grids are being deployed to decrease the cost and make the power systems more reliable. For a Smart Grid to perform effectively, smart devices like smart meters are made use of. The challenge now is to protect the data in these devices from unwanted signals like noise present in the data. Also dynamic state estimation is highly significant for successful implementation of Smart Grid. There can be various approaches for successfully handling this. Kalman filters can be used in smart grids for deriving an optimal performance. This filter is used for identifying the unusual disturbances, device failures and malicious data attacks. Kalman Filter is a dynamic state estimation technique which can do good work for noise variation estimation. This method is able to optimally predict real time data while reducing the problem of nonlinear data like noise. The Phasor Measuring Units (PMU) play an important role in power systems as they are able to monitor the power flow within a network and are made use of for real time measurements. On using PMUs for monitoring, the quality of the smart grid improves. But this comes at a cost, the implementation of PMU invariably enhances the dynamics of noise variance which further leads to an increase in uncertainty in noise-based distribution. In this study the Linear Quadratic Estimation method (LQE) which is the Kalman Filter is used to reduce the amount of uncertainty in noise. The Kalman Filter along with Taylor expansion series is used but this approach consumes a lot of time and is prone to quite a few errors at the time of testing. Due to high complexities of the system it could be hard to derive the process.

The study proposes a technique to work on covariance earlier based estimation using Bayesian analysis where the estimation of the dynamic polynomial prior is done by using Particle Swarm Optimization (PSO). The study compares the results from the PSO optimized Kalman filter and the Kalman filter Covariance Bayesian method.

2.9 Research Objectives

The objectives can be summarized as follows:

- Study of the various parameters of power system which are useful in smart grids.
- The analysis of effects of these parameters in the overall performance of the smart grids
- Solving the problems of Taylor Series expansion and the errors introduced due to linearization using Taylor Series Expansion
- Combining these measurements through Bayesian Filtering for better and accurate estimations
- Estimation of these parameters using the sensors measurement as well as the process model
- Development of non-linear extensions of Kalman Filtering which can handle the non-linearity aspect of our problem

2.10 Outline of Thesis

Chapter 1 gives the introduction, highlighting the concrete energy-based policies to encourage Smart Grid activities around the world. The essential thought of the Smart Grid isn't adequate when relying on the existing intricate framework. Indeed, even with technologies and experiences that are available for reference, the Smart Grid (SG) represents assumption of money, time, and continuous, testing and examination. There are enormous attempts set forth for Smart Grid as this helps in accomplishing ecological protection, safeguarding, and energy maintainability.

Chapter 2 discusses the literature survey conducted on the constituents of the Smart Grid specifically PMU, traces evolution of state estimation and optimization approaches for the Smart Grid to be successful. PMU's play an important role for state estimation, hence in this

LITERATURE SURVEY

extensive literature survey has been discussed on various PMU parameters and state estimation as well as optimization approaches. On the basis of inferences drawn the research objectives were crystalized.

Chapter 3 In this chapter, the Kalman filter along with its extensions are discussed. Kalman filter, Extended Kalman filter and Unscented Kalman filter have been analyzed for parameter estimation, and results compared.

Chapter 4 Kalman filter is a linear quadratic estimation technique. To reduce the non-linearity component in the state estimation, Taylor series expansion was used. To reduce the errors, this was further combined with the Bayesian approach and the results were compared

Chapter 5 To further optimize the state estimation, a hybrid approach along with an optimization technique which is the Particle Swarm Optimization has been discussed.

Chapter 6 The non linear extension of the Kalman filter which is the Extended Kalman filter is explored here for a hybrid approach. Various combinations of techniques have been used and the results analysed.

Chapter 7 The conclusion and future scope of the work is discussed.

2.11 Conclusion

The demand response (DR) will assume a significant form in the SG usage and will permit benefits on framework activity and development on market effectiveness. It will be utilized for reduction of load in light of concerns related to peak power and for ancillary services for frequency-based regulation with fast scale reaction times. This part follows the advancement done in improving the demand response frameworks in SGs alongside parameter estimation of smart-grid advances. PMU plays a significant role in WAMS for PS monitoring. Deployment of PMU has brought a paradigm shift in state estimation.

A brief summary of few optimization approaches has been discussed, which matured over the past decade and basically help to improve the accuracy of data collected in real time to meet the opportunities and challenges for SE presented by the smart grid, within its relevant framework.

In regard of current smart grid development, PMU's are expensive and they enhance the noise dynamics too but at the same time they provide complete observability of the system. Hence, lot of good research on optimal placement of these has been carried out for meeting

LITERATURE SURVEY

various challenges of smart grid-based functionalities present in the system. A brief summary of few approaches which matured over the past decades and basically help to improve the accuracy of state estimation in real time to meet the opportunities and challenges presented by the smart grid, within its relevant framework, have been discussed.

For state estimation ,dynamic estimation of Kalman filters along with PMU data can go a long way in making smart grid more efficient. But since Kalman filter is linear and the noise component being nonlinear, work is required to remove the nonlinear component while predicting the data in real time, hence linearizing and optimizing techniques can be explored to be implemented simultaneously

On the basis of literature studied and research gaps observed and problem formulation the research objectives were arrived at and outline of the work was planned.

Chapter 3   KALMAN FILTER BASED PARAMETER ESTIMATION

It has been stated that for state estimation, dynamic estimation of Kalman filters along with PMU data can go a long way in making smart grid more efficient. In this chapter, the popular Kalman filter and its extensions are discussed.

3.1 Introduction

In this modern era of structural monitoring for health, control, validation of model, detection of damage, recreating the feedback of the already existing structure, is all partially subjected to anonymous type of excitations . If the system structure is not of instrumented form, the challenging part is the estimation of the unknown type of loads and to further implement all of these in sync with the structure model and finally to obtain the response of the system. In such a methodology, the estimation-based accuracy depends on how accurate the model is and how it is predicted and the type of loading to perform the system physically. Though, if in case the system structure is of instrumented form and response measurements at similar places are available, the estimation accuracy gets enhanced by the combination of the model and the measurements mostly in a logical manner. The response-based remodelling idea for specific system of instrumentation has been analysed by the process of state estimation and area of control since the early 1950s, and mostly, new developments and contributions began to appear in the corresponding years.

This operational work generally focussed over some of the applications of theory based on Kalman filter (KF). The KF formulation generally consists of a model blended with various type of measurements. The improvements on the basis of response-based predictions, comparative to open loop system, usually results from incorporation of the data measured into process of estimation. The "open loop" represents a system which does not involve any feedback in terms of reaction from the system output, which is opposite to a system having a closed loop, where the excitation-part in general depends on the system feedback in terms of the response obtained from the system. The theory of Kalman filter has received a challenging scrutiny in most of the fields like economics, engineering (electrical), navigation, and robotics since the early 1960's. The applications and the complex problems commenced the use and development of Kalman Filter theory. Moreover, the use of Kalman Filter process

KALMAN FILTER BASED PARAMETER ESTIMATION

becomes more outstanding due to the certain issues that may utilize it more commonly than ever before, and there consists a potential key and merit in incorporation of KF theory into the existing form of structural-based systems.

In regard to critical and complex existing infrastructures generally new challenges arise, mixing element models (finite) and the data measured presents a potential to become a progressively significant part of a research for structure-based engineering [36]. The dynamic state of the system represents the variables that generally provide a representation (in whole) of the internal operating condition or system status at a known time instant. If the state of the system is given, then the system evolution can be predicted easily, if the system excitations are provided (known). In another manner one can say that the system state involves certain variables that help to specify the primary condition of the system. When the system-based structure model is applicable or available, the dynamic behaviour of the system can be observed for a known system input by generalizing the equations related to the motion of the system. On the other hand, if the system structure is exposed to disturbances of unknown type and if the system is instrumented partially, the response generated at boundless degrees of freedom is achieved with the help of state estimation process. The primary idea in theory of state estimation is to achieve approximations/likeness of the system state that is not observable directly by the use of model-based information and from any of the measurements available. There are basically three of the estimation issues stated as follows:

- Estimation of the present state, $_n$ with the given data (in all) that also involves y_n; this process represents filtering.

- Estimation of the prior (past) value of x_k, where k < n with the given data (in all) that also involves y_n; this process represents smoothing.

- Estimation of the future value of x_k, where k > n with the given data (in all) that also involves y_n; this process represents prediction.

The process of state estimation is generally observed as a mixture of informational data from the analytical form of system model and the measurements form making certain predictions about the states i.e. both the current and upcoming future states of the system. The process of uncertainty in sate estimating (dynamic) process generally arises from three of the following sources:

- Stochastic type of disturbances.

- Inconsistency between the model and the practical system is generally used to represent the uncertainty of the model.

- Unknown primary operating conditions.

The state estimation theory begins from the methods of least squares that established essentially in early 1970s by the work [101] conducted by H.W Soreson (1970). The mathematical kind of framework in terms of current state estimation theory originated by the work of Wiener in early1940s. This field gained its maturity in the late 1960's and 1970's after a discovery led by R. E. Kalman (1960) that was popularly known as the Kalman filter or in short KF. The concept of KF presents a recursive form of data processing algorithm that provides an optimized estimation of system subjected to the stochastic stationary disturbances with well-known covariance.

The famous scenario of loads that are electronically controlled and the large use of alternative form of energy sources like solar cells and wind turbines, the quality of power at distributed level should be monitored carefully. The different means for monitoring the quality of power is through the measurements of frequency [20]. The nonlinearities of the system cause effective harmonics generation that distorts the signals observed at the load level, and so such an easy task also becomes a major challenge. The recent developments in computing digitally carries strong linearized estimators such as the Kalman filter to be implemented widely for the estimation of frequency. In practical applications, the processing systems usually exhibit non-linearized behaviour that eventually raises the desire for adapting the KF in order to fit non-linearized models. The EKF (Extended Kalman Filter) generally deals with the system-based nonlinearities with the help of model-based linearization around the state that is known. The recent development of UKF (Unscented Kalman filter) takes the benefit of UT (Unscented Transformation) in order to face the models with nonlinearities without any kind of linearization and with similar computation-based complexity as in case of EKF. However, in case of modelling an uncertainty during the EKF, which is very sensitive to initial operating conditions representing the limitations of the system, an upgraded form of UKF algorithm on the basis of STF i.e. strong tracking filters has been developed. If the algorithm is properly tuned, then it helps to improve the tracking for unexpected transformations and it further avoids the property of divergence. However, if the measurement and/or processing

KALMAN FILTER BASED PARAMETER ESTIMATION

noise gets converted or gets changed along with time, then such a filter will not be able to estimate with similar kind of accuracy as it was tuned earlier. UKF following an [45] adaptive algorithm is generally meant to boost the estimation frequency of energy signals that undergo various changes and are damaged by white noise. The adaptive algorithm is generally based on the configuration of Master-Slave, where the system "master" helps to estimate the system state and the "slave" of the configuration, operates in parallel, and estimates the noise-based covariance matrix. As the signal voltage is distorted, the distorted signal is generally employed for deriving a complex model of state-space, which estimates the frequency on fundamental basis.

3.1.1 Kalman Filter

The combination of certain methods of dynamic estimation of state like the Kalman filters with practical (real) time-based data collected by the digital type of meters likewise the phasor measurement units (PMUs) can lead to progressive form of techniques that improves the quality of controllability and monitoring in the technology of smart grids. The Kalman filters (KF) of classic form generally achieves the optimized performance with ideal model of the system that is really hard to achieve practically with sudden type of device failures, disturbances, and the data attacks of malicious form. In such type of work, one usually introduces and compares a method of novel type, viz. adaptive form of KF with inflatable form of noise-based variances, against a collection of classic KFs.

In recent years, the concept of power system has become very dynamic and complex. While the estimators of the state use the data measured from SCADA i.e. supervisory control and data acquisition system in the time-based intervals of few seconds, such a level in its conventional form is not translated or converted into methods for properly capturing the dynamics of the system. Over the past few times, the concept of measuring the phasor technology guarantees key advancement on the basis of real-time system state tracking, specifically when joined with the techniques of dynamic state estimation. In conventional studies of power system, the state of the system can be assumed by a joint combination of KFs with abstract measurements available and true model. In such an ideal methodology, the system model in its hypothetical manner contains proper characteristics on the basis of noise statistics integrated into it. In real knowledge, the statistics of noise is not known properly in reality. In case of modelling, the system-based measurements are also corrupted with some errors due to malicious attacks done to data, or even failure of the device. In order to carry the activities of illegal form such as energy theft, the attackers of the system, design or model a bad type data that bypasses the mechanisms of detecting bad data in the power systems. This evolves certain overwhelming impacts such as wrong rustle (dispatch) in the case of power system distribution and the breakdown of device during the generation of power. The Kalman filters (KFs) are very useful in the applications of real world involving: robotics, GPS, communication systems, chemical-based plant control, weather prediction, aircraft tracking, satellites, inertial navigation, ships, multi-sensor based data fusion, people, rockets, cars, and cows, in addition it includes some of the reluctant applications (for example, prediction of stock market). However, the Kalman filters (KFs) are easy to code and model, and they regularly help to provide accuracy in estimation process. Differently, the accuracy of Kalman filter can be exceptionally bad for most of the real applications, for various reasons, consisting of: 1) equation-based nonlinearities describing the system physical state, 2) covariance matrix ill-conditioning process, and 3) incomplete or inaccurate models of pointing out the problems in physical form. Basically, a great section/part of the algorithms related to dynamic state estimation are based on Kalman-filter (KF).

Kalman filter is an optimized estimator as it estimates the error covariance of the system and predicts recursively for timely system-based measurements. The distinct property of KF can be categorized into two distinct processes. The first one is the process of prediction and the second one is the process of measurement. The two processes are together driven in a recursive form for optimized action of Kalman filtering. In its recursive manner, the realization of KF is usually explained as an implementation in real time.

Advantages of using Kalman Filter

- The algorithm based on Kalman Filter can be implemented over digital computers, which was replaced by the circuitary of analog type both for the control and estimation process.

- For random processes or deterministic dynamics, the stationary properties of KF are not needed.

- The KF is consistent with state-space formulation of optmised controllers for dynamic systems. It is fruitful for both the estimation and control properties of the system.

- The KF needs less amount of mathematical preparation.

3.1.2 Ensemble Kalman Filter (KF)

In practical systems, the relationship of measurement and process to the state are often non linear and one of the better and possible solution is the EKF i.e. Extended Kalman Filter, which linearizes the functions that are of non-linear nature that lies around the interest point named 'x' in state space. The concept of EKF [84] developed for use in applications like aerospace and motion, where the state space is small. However in case of systems with a state space i.e. high-dimensional like in case of disease and weather prediction, it fails due to stubborn computing burden linked with matrices such as Jacobian matrices, a learning time-based step. One such example to overcome such kind of deep labour intensive evaluation is to use the techniques of data catabolism, like the particle filter and ensemble Kalman filter (EnKF). Such type of filters are generally related with the sense that each of the particle acts as a member of ensemble; the main difference is that the concept of EnKF comes up with an assumption of the characteristics with Gaussian distributed noise, and in that case, the EnKF is considered to be more active than particle filter. The important ideas of EnKF include; (1) maintenance of state estimates ensemble that represents state samples collection rather than an individual estimate, (2) simple advancement of each of the ensemble member and (3) evaluating the error covariance matrix and the mean on direct basis from such an ensemble rather than maintaining covariance matrix separately. In terms of comparison, the gain of Kalman is generally calculated as the following:

$$K_k = \frac{corr\left(\hat{x}_{\bar{k}}, \hat{z}_{\bar{k}}\right)}{cov\left(\hat{z}_{\bar{k}}\right) + cov\left(z_k\right)} \ldots \ldots \ldots \ldots (2)$$

Where, the *cov* and *corr* represent the *sample covariance and* the *sample correlation* operations respectively as presented. The ensemble KF greatly offers an easy implementation process and handles the non-linearity because of Jacobian calculations absence, on the other part, it is complex to choose the size of an ensemble, which is considered to be of statistically representative form normally large enough for the process. The selection of optimized size of an ensemble must be addressed on the basis of trade-off knowledge between complexity of computation and estimation accuracy.

3.1.3 Inflatable Noise Variances (In NoVa) with Adaptive Kalman Filter

Keeping in mind that the model of state evolution may be incorrect often and the data of the system is more prone to man-made and instrument error. An individual is persuaded to develop such kind of methodology, AKF with Inflatable Noise Variances (In NoVa). By definition, such type of errors are unpredictable and of unknown type, such that they are not reflected in the process of noise covariance (Q) and the measurement-based noise covariance (R). The concept of AKF with In NoVa usually differs from the mentioned (above) techniques of KF and in such a case it dynamically adjusts with Q and R by the process of analysing the residual and innovation. The term residual is generally defined as: $I_k = (z_k - H\hat{x}_k)$ that represents the difference between the quantity of measurement and that being evaluated from *a posteriori* estimate of state. The term innovation on the other hand is generally defined as: $I_k^- = (z_k - H\hat{x}_k^-)$ that represents the difference between quantity of measurement and that being evaluated from *a priori* estimate of state. On ideal basis, the term innovation must be distributed normally with covariance and zero mean: $S_k = HP_k^- H^T + R$; and on similar basis, the term residual must be also distributed normally with covariance and normally: $T_k = RS_k^{-1}R$. The concept of naive RKF is capable in a restricted manner, of identifying and detecting the data of bad form. Under the conditions of abnormality, the distribution of innovation will transform and the RKF becomes capable to capture it. This happens when the innovation of normalized form and is normalized by its S_k covariance exceeding a predetermined form of threshold. Moreover, since the R and Q already get mixes in S_k, it is really not possible to determine whether the ambiguity is caused by false processing model, a false type of measurement, or by both .However, the concept of AKF with In NoVa, by investigation of both the residual and innovation, can possibly be able to distinguish the measurement errors from process errors.

3.1.4 Extended Kalman Filter

The Extended Kalman filter, popularly known as EKF was developed for non-linearized discrete time-based processes. It provides an optimal estimate of approximation. The nonlinearities of system-based dynamics are usually estimated in approx. by the linear version of non-linearized system model around the estimation of the last state. For a valid type of approximation, this process of linearization using the first order Taylor series should be presented as a good approximation of non-linearized model in global uncertain domain linked with the estimation of state. The concept of EKFs faces various challenges from the use of linearization processes like challenges of implementation, tuning difficulty, issues related to reliability, etc.

KALMAN FILTER BASED PARAMETER ESTIMATION

KF is suitable only for the systems that are of linear nature, and it generally needs to have a linear form of observation equation. In practice, most of applications represents the non-linear systems. Therefore, the methodology of nonlinear filter is very significant. In the field of non-linear filter, the most classic form of algorithms is EKF. The basic process of EKF is to capture the value of 1st order Taylor expansion around estimated states and then to change the non-linear system into an equation of linear type. The EKF algorithm is used often for non-linear filtering systems, and eventually simplifies implementation. Moreover, Taylor expansion usually belongs to a process that is of linear form, so if the observation equations and the system status are nearby continuous and of the linear form, then the obtained results of extended Kalman Filter are nearly close to the true value. Additionally, the result based on filtering gets affected by the noise measurement and status of the system. The system-based covariance matrix status and observational noise generated by the system remains unchanged in the methodology of EKF. If the noise-based covariance matrix of observation and status are not accurately estimated, then the error of cumulative form may result in the process of filter divergence.

The estimation-based on EKF is capable of estimating two of the dynamic states of the system . In its first step, EKF is usually premised on basis of 1st order expansion of Taylor series for equations of nonlinear type. The results obtained from EKF would greatly differ from the original value, if the extended Kalman Filter were used in the case of expansion point or nonlinear systems deviated greatly from original value. Secondly, the EKF generally assumes that the noise based on the observation and status is independent in case of processing of white noise, but the noise-based characteristics sometimes are not able to meet the requirements of white noise characteristics. As the status of the system and observational noise may differ after 1st order expansion of Taylor series, the noise-based assumption will also be inconsistent on real basis. EKF requires recalculation of Jacobian-based matrix of current (present) time observational equation as in case of observational time for each of the processes based on EKF.

The Jacobian calculation yields a model function trajectory centred on the location of state. The formulation is represented as follows:

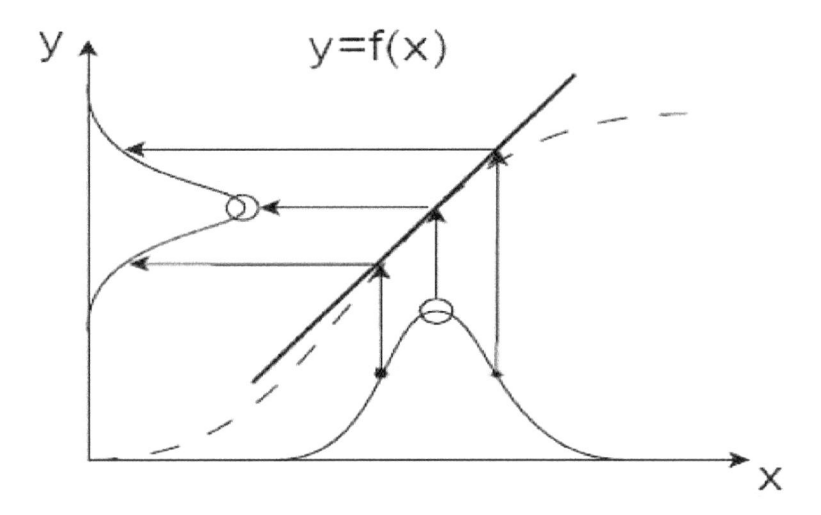

Figure 3.1 EKF linearizing non-linear function around mean Gaussian distribution [55]

But the matrix-based calculation is very hard and complex. So, in most of the cases, it is really hard to analyze the Jacobian matrix.

The Kalman filter (KF) theory is generally applied to the problem of linear-Gaussian, but most significant applications of real world are non-Gaussian and/or nonlinear. Engineers generally use the linearized approximations in order to use this theory for the non-linearized issues, which are common in the applications of real world. Such kind of linearized approximations are very easy to use thus explaining the demand of EKF. An issue related to linear filter represents all the functions i.e. f, G, and h, which are linearized in x, and a "Gaussian" problem represents all the noise (v and w) which are Gaussian in nature. The extended KF approximate (f, G, and h) the Taylor series (first-order) are observed at the current (present) estimate. Particularly, distinct EKFs can be modelled by using a large list of tricks-based on engineering such as different coordinate systems, different covariance matrix-based factorizations, higher order (or second order) corrections of Taylor series to the prediction of state vector and/or the update of measurement, state vector iteration updated by the use of measurements, distinct orders to use the sequential form of scalar value measurements in order to update the vector state, quasi-decoupling, tuning noise process and combinational joint effort of all, as well as the other whistles and bells discovered by analysts engineers in hope of enhancing the performance of EKF. In some cases the practical form of tricks generally result in wide advancement. The concept of distinct system coordinate for example, the Polar and the Cartesian may result in significantly distinct EKF accuracy, that sometimes creates trouble for both the mathematicians and physicists, as the "physics" does not depend over what the system-based coordinate is used for describing it, and the experts of mathematics generally feel that the descriptions that are "coordinate free" are simple, more powerful and general. The non-linear and linear filter issues have the similar theoretical optimized accuracy of estimation in any of the coordinate system. Though, the engineering experts know that the accuracy of EKF may depend upon the specific used coordinate system, owing to ill-conditioning and/or nonlinearities. However, there is no such law that is build up against the use of more than a single coordinate system in a known extended Kalman Filter. Such coordinate systems of hybrid form have been used in EKFs with a good impact in case of radars for past decades.

The EKF represents the main methodology that is used for estimating the state of power system on current basis. But there are few of the drawbacks: The Linearization of continuous discrete (CD) will cause filtering instability if the time interval of calculation is not limited enough.

- For complex systems and large-scale systems, the calculation of Jacobian matrix is considered to be complicated that is to be executed in case of real time.
- The disregard of higher and quadratic order terms makes estimation and the prediction less correct/precise.

3.1.5 Unscented Kalman Filter

Unlike the EKF, Unscented Kalman Filter (UKF) uses complex linear approximation. But instead the UKF uses an accurate kind of approximation in order to evaluate the integrals of multidimensional form desired by the theory. Such an approximation is usually known as the transformation of unscented type, for most of the genuine reasons, it is similar to the quadrature of Gauss-Hermite form for the approximation (numerical) of multi-dimensional integrals [84]. The UKF generally assumes the state vector probable density representing the Gaussian form, and this kind of density is usually sampled carefully at selected points in order to approximate the required multi-dimensional form of integrals.

Specifically, the EKF helps in approximating the expected $f(x)$ value as $E[f(x)] ¼ f E(x)]$, that represents the exactness for the linear type of functions, but otherwise, it may be very bad kind of approximation. For instance, considering a simple form of nonlinear functions $f(x) = x2$. Suppose that the mean of x is considered to be zero i.e. $E(x) = 0$, resulting in approximation of EKF i.e. $E(x2) ¼ [E(x)] 2 = 0$. However, the accurate or exact $E(x2)$ value may represent any positive number i.e. the variance of x. The linear approximation of simple type is used by EKF in such a case that may be wrong by great amount. In contrast, the approximation by UKF [83] is better than EKF, and the computing complexity of the both EKF and UKF are mostly similar. The UKF represents a method of non-linear filter on the basis of unscented type of transformation in order to evaluate the probability of the random non-linear vectors. It involves a set of points, where the variance is P_x and the mean value is $\hat{x}$ in order to get variance P_z and measurement mean $\hat{z}$ with the help of non-linearized function based on state transition and the function of measurement.

Such kind of method helps in decrementing in computation complexities of the system and at the same time boosting the accuracy of estimation, having faster and more accurate form of results. The approach of UT helps in providing another noise benefit in a non-linear fashion for accounting for non-additive or non-Gaussian noise. For performing such methods, the noise gets propagated through functions by state vector augmentation including the various sources of noise. Such a method was first provided by Julier first and then later on it was developed by Merwe. The samples of Sigma point are then chosen from the state augmented x_a including the values of noise. Such a technique mainly results in capturing measurement noise and process-based accuracy with similar accuracy as in case of state which in return enhances the accuracy for sources of non-additive noise.

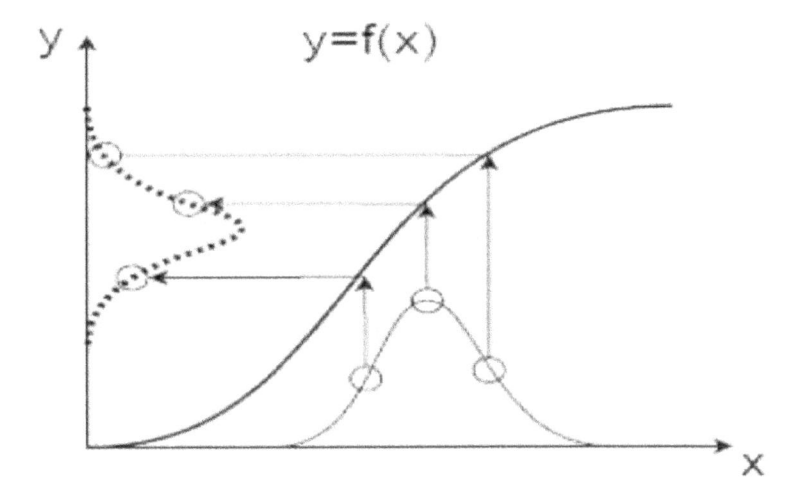

Figure 3.2 UKF based propagation of sigma points from the distribution of Gaussian form through a functions of non-linearized form [55].

On comparing with the filter of extended form or EKF, UKF is simple to implement, and it helps to avoid Hessian and Jacobian matrices calculation. Wang, H., et al (2013) applied [45] the UKF for analyzing the tracking of flying target for upgradation of target-based prediction matrix. The other target prediction matrix state is generally used as the trajectory-based prediction value. The experimental results of processed data show that the UKF provides a large convenience for processing on real-time basis. Rajagopalan, A. N. et al (2012) proposed the UKF formulation estimation of depth, and remodeled the scene-based 3D structure from optical defocus/ motion blur [51]. Gao, J. et al (2012) overviewed a UKF algorithm on the basis of technology based on electronics. Simulation-based experimental results indicated a variable form of accelerated targeted motion that was traced under 3D coordinate system showing that the improved form of algorithm helps in achieving good robustness and precision [52]. Han, J. et al (2012) implemented the adaptive form of UKF for the purpose of visual tracking, and an improved version of tracking based on accuracy and real-time [53]. Seok Han Lee jointly combined the UKF with particle filter for camera tracking on the basis of real-time and it further concluded that the UKF-based tracking helps to grab successful feature mapping and camera tracking in a real environment.

The analysts as an application have used the state estimation on dynamic basis presenting a state estimation branch. On actual basis, the power system represents a non-linear, dynamic, and complex system. State estimation in dynamic form is in line with power system nature than that of the static-based state estimation process. The process of dynamic state estimation that involves the capabilities of forecasting may help to check the power system operational status on the basis of real-time. So, it represents a key area of the energy-based management system. Presently, the method of dynamic state estimation of the power system is generally based on EKF methodology. In case of normal suitable conditions, comparatively it is very much accurate to use the method of EKF in order to obtain the dynamic state estimation of power system. Whereas, in most of the cases like generator or load power output mutates (variate), the restriction of EKF methodology which results in production of large error. The Adaptive Kalman filtering (AKF) is generally used in improving the accuracy of the filtering process. Because of its online-based parameters of estimate model and statistical-based noise characteristics, the amount calculated is very large and it was very difficult in meeting the online specific needs. UKF approach is used for dynamic state estimation, and is more accurate in estimation process than the conventional EKF. But the UKF concept has limitations too. It can only be used for temporary form of Gaussian-based distribution model. While the original form of power system represents non-linear type of system, specifically after huge disturbances of the system. The load will be transformed and the generators on virtual basis have large amount of oscillations. Such kind of transformation and oscillations are of nonlinear (highly) nature, and the overall system represents a nonlinear time variant system, using the method of UKF for dynamic form of state estimation can lead to few possible defects.

3.2 Observational Comparative Performance

The utilities of electric energy are becoming more anxious about the voltage distortion and harmonics of power system. This rising concern occurs due to rise in power electronic devices application occurring in all the operation like inverters and rectifiers. This process results in increased power system harmonics occurring in the system. Repeatedly, due to increasing application of system-based series and shunt capacitors, the static var controllers (SVCs) are located at strategic locations for the correction of power factor. For the conditions of resonant type, there are wide chances of rising potential that magnifies the already existing levels of harmonic. The components of power system regularly inject harmonics that is of time varying nature giving rise to harmonic currents and voltages (non-stationary) in case of distribution system. All the functions in real time are of a non-linear nature and these systems can be presented as accurate discrete time-based systems to a large extent in small time intervals. Currently, the main issue is to provide state estimation of discrete-time controlling process and such type of processes is usually conveyed using stochastic (linear) equation. This estimation is done accurately and easily using Kalman Filter. But for nonlinear measurement and processing systems then the EKF and UKF are generally implemented . A Kalman Filter, which linearizes about the present form of covariance and mean using any form of linearized function, which is known as extended KF (EKF). In this part the measurement functions and the partial form of derivatives are used in order to evaluate estimates in presence of non-linearized functions.

The performance of algorithms , both the EKF as well as the UKF in general depends over the measurement of data quality. For instance, the nature and the content of noise in both , availability and measurement of data, for example, irregularity in measurement and the rate of updating the measurement data. Nishiya et.al. (1982) studied and presented [97] an algorithm based on KF for estimation of dynamic state in order to estimate bus angles and voltages that generally had bad data-based anomalies, change in topology of the network as well as the variation of sudden form in various states of the network. Estimation on the basis of real time using the method of robust extended Kalman Filter (REKF), which performs better than the EKF, is generally proposed for power system-based estimation of harmonic states . With less amount of measurements number than the EKF, the robust EKF (REKF) has better capability of estimation under the condition of bad data with the use of IEEE 14 bus system . In case of system-based on WSCC 3- generator 9-bus test and single machine infinite Bus (SMIB), the analysis of EKF on the basis of estimator under the operating condition of three-phase to ground fault and sudden change of load is generally analyzed with peculiar conditions of measurement. The use of improved version of EKF with the help of Euler method based on second order for the extended Kalman Filter with prediction of multi-step is usually offered and the performance-based analysis in case of parametric error, composite error, and topological error has been established with the measurement type of update 0.04 s interval with the use of 16- machine 68-bus system as referenced in [20]. Moreover, [20] indicates a possible way of analyzing the EKF performance in case of unavailability of measurement data. Kamwa and Ghahremani have proposed an algorithm EKF with unknown inputs (EKF-UI) in order to estimate the unknown inputs and the dynamic states for system based on SMIB [58]. The UKF that performs well as compared to EKF and WLS (Weighted Least Squares), for three of the distinct test standard of the system, is presented agreeably for the case of distinct measurement and transient operating conditions. For the case of SMIB and WSCC (Western System Coordinating Council) system, UKF is proven to perform better estimation process than the estimator EKF for measurement type of data having distinct noise-based content with update intervals of measurement data.

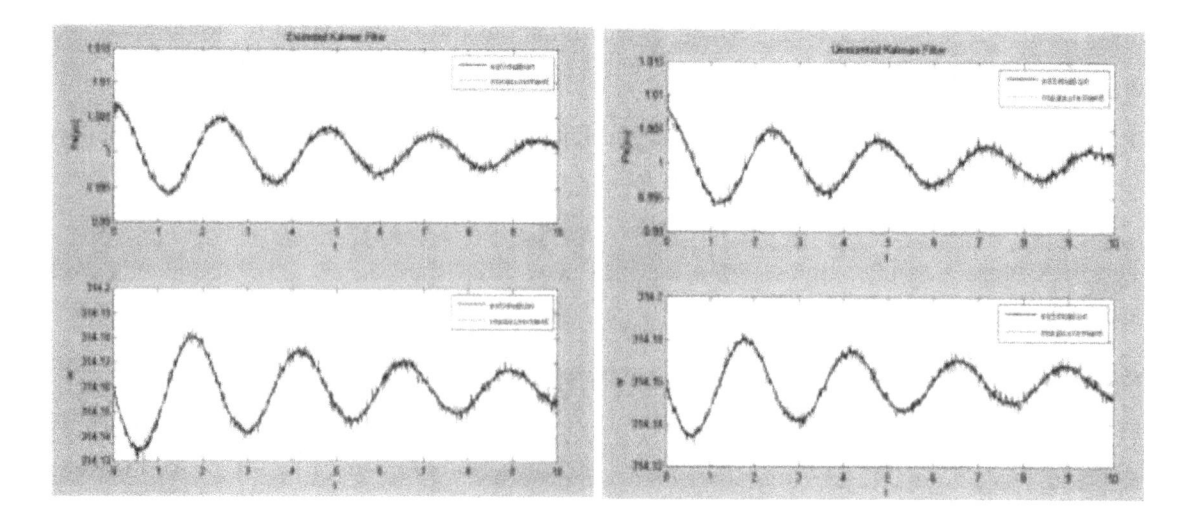

Figure 3.3 EKF and UKF for Single machine system

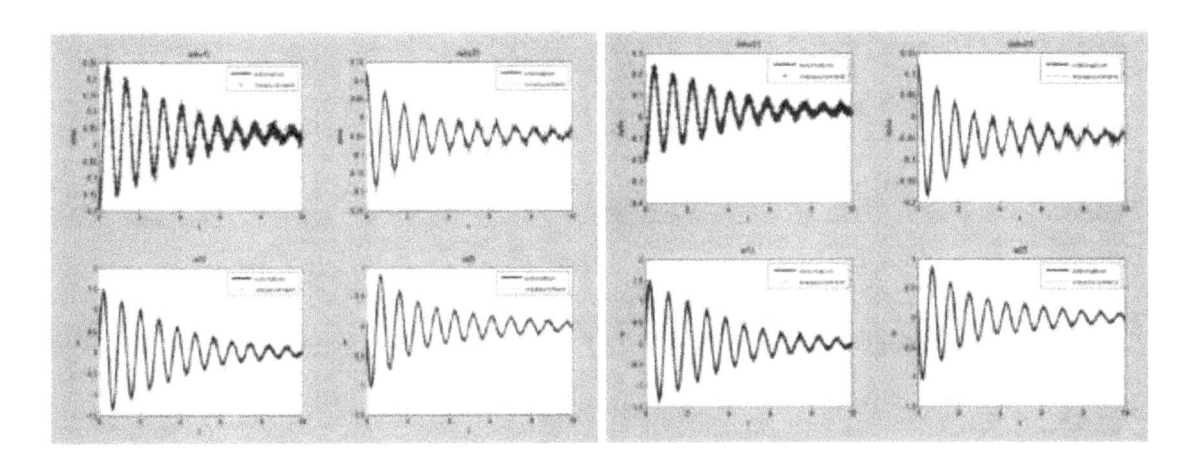

Figure 3.4 EKF and UKF for Multi machine system

3.3 IEEE 30 Bus System

The system considered is the IEEE 30-bus system. The system is characterized by 15 buses, 2 generators, and 3 synchronous condensers. The 11 kV and 1.0 kV base voltages are guesses, and may not reflect the actual data. The model actually has these buses at either 132 or 33 kV; what is worth mentioning is that the 30-bus test case does not have line limits.

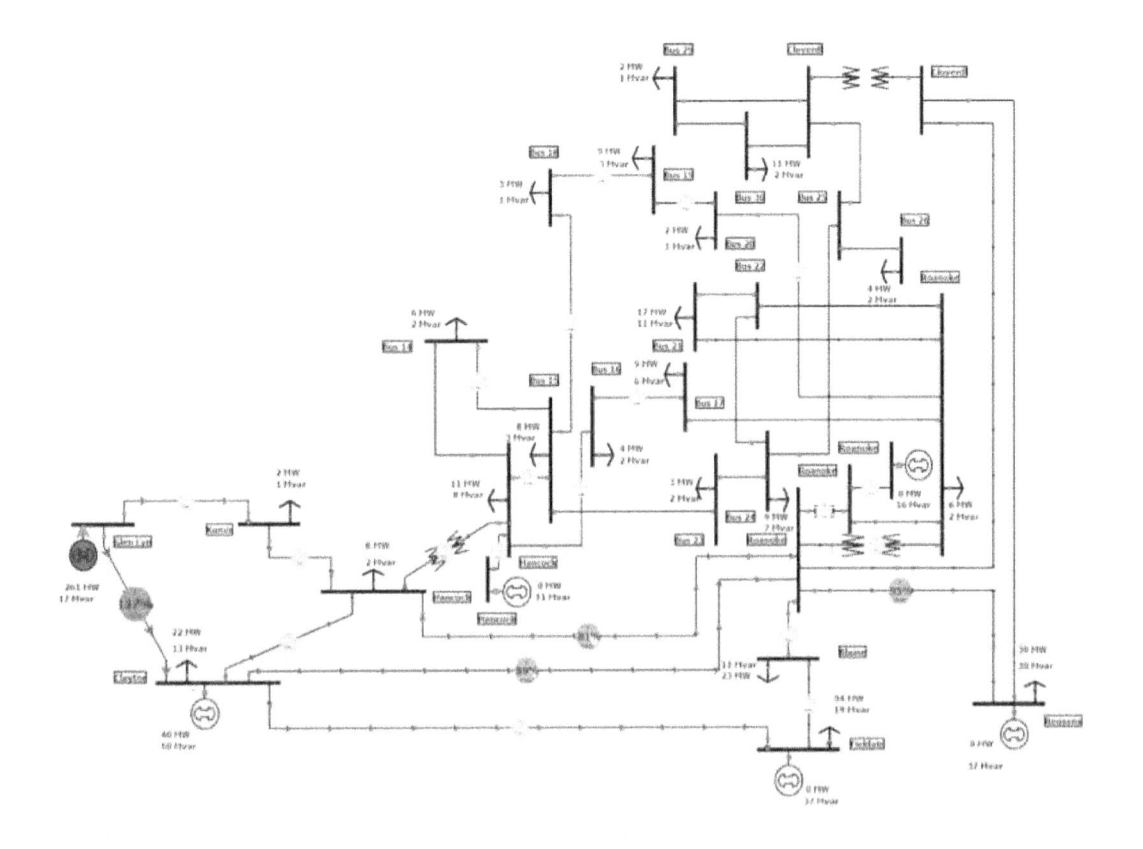

Figure 3.5 IEEE 30 BUS System

- Number of PMU's used 20
- Location of PMU depend on Noise value greater than filter Threshold

3.4 Performance of Kalman Filter

It was proposed as a recursive form of squared mean error minimizer in order to estimate the processing state, using the measurements of noisy form considered as feedback. The equations of the system falls into two of the following groups, "Predict" (time update) and "Correct" (measurement update). These two classes are illustrated as follows:

$$\text{Predict} \begin{cases} \hat{x}_{\bar{k}} = A\hat{x}_{k-1} + Bu_{k-1} \\ P_{\bar{k}} = AP_{k-1}A^T + Q \end{cases} \quad \text{...... 1(a)}$$

$$\text{Correct} \begin{cases} \hat{x}_{\bar{k}} = \hat{x}_{\bar{k}} + K_k(z_k - H\hat{x}_{\bar{k}}) \\ P_k = (1 - K_k H) P_{\bar{k}} \end{cases} \quad \text{...... 1(b)}$$

Where, $\hat{x}_{\bar{k}} \varepsilon R^n$ = a priori estimation of state at 'k' time step given with process-based knowledge prior to 'k'.

$P_{\bar{k}}$ = a priori covariance error estimate.

Q = process-based noise covariance.

The state of prediction helps to project x_{k-1}(previous state) and x_{k-1} (error covariance) i.e. forward on time basis in order to obtain the estimates of a priori for 'k' time step.

$\hat{x}_{\bar{k}}$ = a posteriori state-based estimation at 'k' time step given with measurement-based vector

$z_k \varepsilon R^m$.

P_k= a posteriori covariance of estimate error.

R= measurement-based noise covariance.

The correction involves some new measurements into the state estimation through the Kalman gain matrix (K) that calculates or weighs that how much z_k is being trusted vs. the prediction of the system $\hat{z}_{\bar{k}} = H\hat{x}_{\bar{k}}$. The larger the value of R, the more an individual trusts the value predicted; the smaller the value of R, the more an individual trusts the values measured.

KALMAN FILTER BASED PARAMETER ESTIMATION

The process model constraints help to make the problem of estimation easier. The Kalman filter constraints in terms of both f and h functions are considered to be of linear form, the terms of noise v and w are uncorrelated, zero mean on the basis of white and Gaussian forms. As the model is of linear form and the input represents Gaussian, then it states that the output will also be of Gaussian type. Therefore, the output and state will be always distributed normally and knowledge of both the covariance and the mean will be satisfied. The estimation based on KF is simple as it involves approximately all the calculation of linear form except the process of matrix inversion.

Kalman filter was applied to the IEEE 30 bus system, discussed above and the changes in the voltage level and noise was studied for different time frames.

Figure 3.6 depicts changes in noise. The parameter is monitored and results are based on filters performances for the time changes in days.

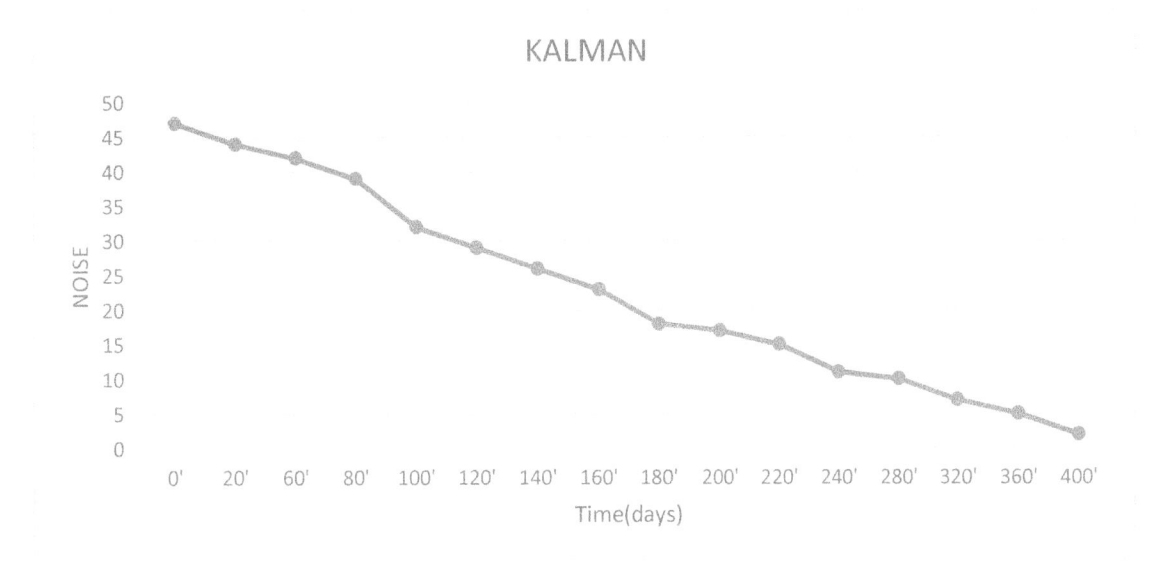

Figure 3.6 Kalman filter response on noise during different time frames

In study of the Kalman filter in fig 3.6, it is seen that in the beginning, the noise is around 47 units per kunit of the measured parameter but continuously decreases with the increase in time and at 400 days, settles down to around 3units, which shows that noise reduces for Kalman filter as time increases.

Figure 3.7 depicts changes in the voltage magnitude. The parameter is monitored and results are based on filters performances for the time changes in days.

KALMAN FILTER BASED PARAMETER ESTIMATION

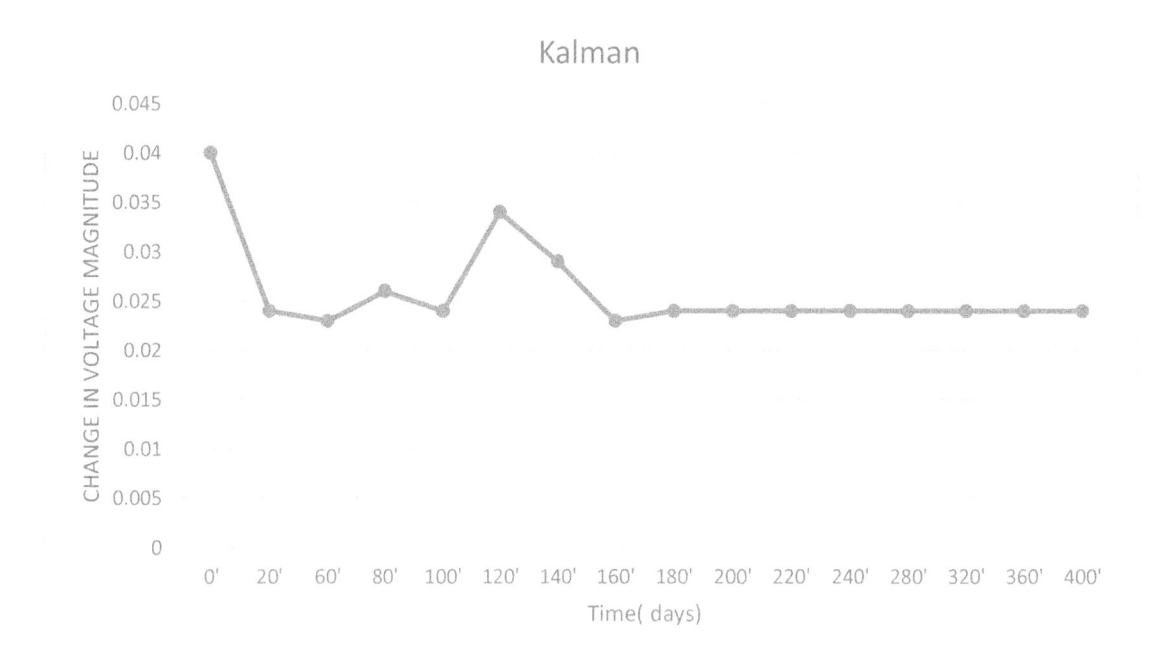

Figure 3.7 Kalman filter response on change in Voltage magnitude during different time frames

The figure 3.7 depicts changes or errors in voltage magnitude where as it should be stable. At starting, there is a wide variation, varies from 0.04 to 0.024, hence, not useful parameter estimation, but at around 180 days, parameters find effective filtering and voltage magnitude becomes stable.

3.5 Performance of Extended Kalman Filter

As already stated, for the EKFs, it is not essential that the observation models and the transition state are to be linear functions of the state but may instead be differentiable functions.

The equations of the system are presented as follows:

State equation: $x_{k+1} = f(x_k, u_k) + w_k$ ……………..3 (a)

Observation equation: $y_k = h(x_k) + v_k$ ……………..3 (b)

Where,

h = predicted measurement from the predicted state.

f = predicted state from the previous estimate.

However, both the h and f are not applicable to the covariance on direct basis. Rather than the partial derivatives, (the Jacobian) matrix is normally computed. For each step of time, the Jacobian matrix is computed with present predicted form of states. Such type of matrices are used in the equations of Kalman filter. This type of process necessarily helps to linearize both the current estimate and the non-linearized function. The Kalman filter (KF) represents a method of linear state estimation. On the other hand, the power systems is of extremely non-linear nature. For applying the KF to original state estimation of power system, an individual needs to perform system linearization. The EKF is generally based on the process of linearizing the non-linear type of equation with the use of Taylor Series where higher order and quadratic terms are erased. The algorithm in detail is presented by the following equations:

$$\text{Prediction: } \hat{x}_{k+1} = f(\hat{x}_k^+) + q_k \quad \ldots \ldots (4)$$

$$P_{k+1}^- = F_k P_k^+ F_k^T + Q_k \quad \ldots \ldots (5)$$

$$\text{Correction: } K_{k+1} = P_{k+1}^- H_{k+1}^T (H_{k+1} P_{k+1}^- H_{k+1}^T + R_{k+1})^{-1} \quad \ldots \ldots (6)$$

$$\hat{x}_{k+1} = \hat{x}_{k+1}^- K_{k+1}(z_{k+1} - h(\hat{x}_{k+1}^-)) \quad \ldots \ldots (7)$$

$$P_{k+1} = (1 - K_{k+1} H_{k+1}) P_{k+1}^- \quad \ldots \ldots (8)$$

$$F_k = \frac{\delta f(x_k)}{\delta x_k}, H_{k+1} = \frac{\delta h(x_{k+1})}{\delta x_{k+1}}$$

By evaluating the Jacobian functions *f* and *h* around the state estimated, problem of non-linearity is generally solved using EKF.

Extended Kalman filter was applied to the IEEE 30 bus system, discussed above and the changes in the voltage level and noise was studied for different time frames.

Figure 3.8 depicts changes in noise. The parameter is monitored and results are based on filters performances for the time changes in days.

KALMAN FILTER BASED PARAMETER ESTIMATION

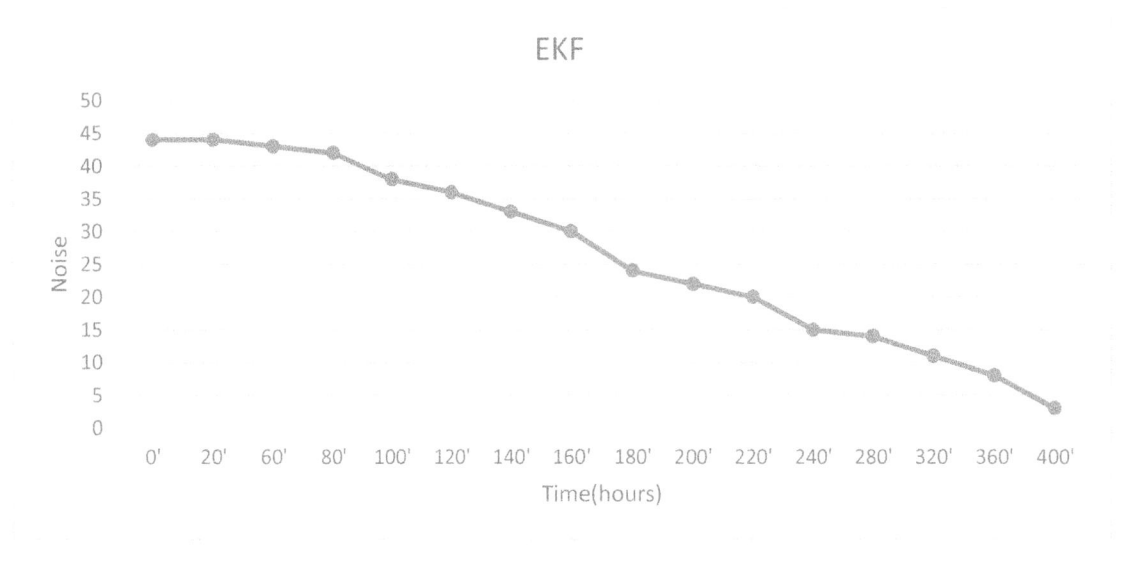

Figure 3.8 Extended Kalman filter response on noise during different time frames

In study of the Extended Kalman filter in fig 3.8, it is seen that in the beginning, the noise is around 44 units per kunit of the measured parameter but continuously decreases with the increase in time and at 400 days, settles down to around 3 units, which shows that noise reduces for Extended Kalman filter as time increases.

Figure 3.9 depicts changes in the voltage magnitude. The parameter is monitored and results are based on filters performances for the time changes in days.

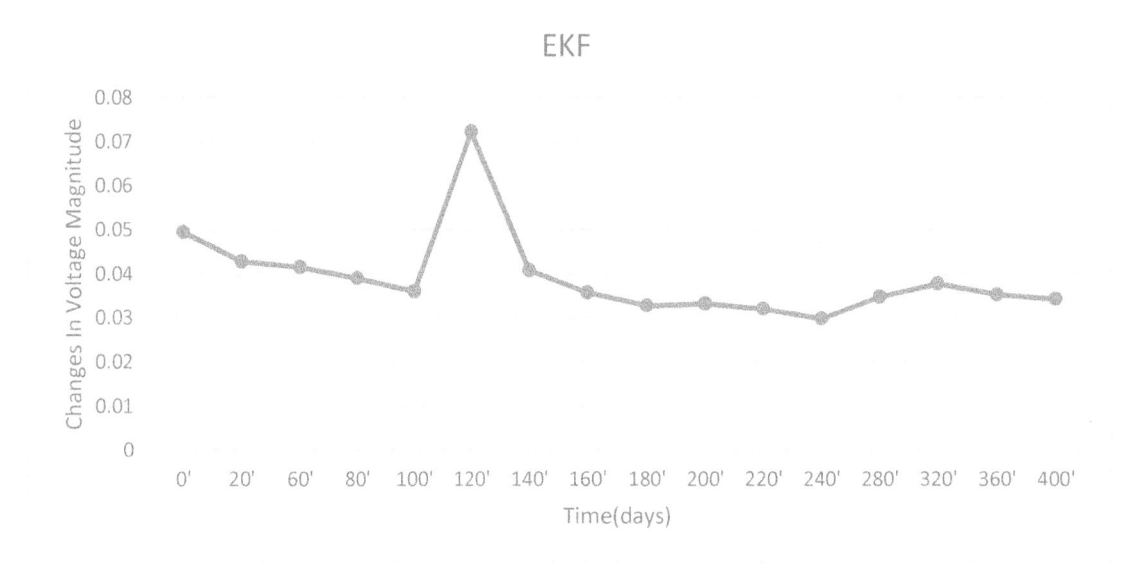

Figure 3.9 Extended Kalman filter response on Voltage magnitude during different time frames

KALMAN FILTER BASED PARAMETER ESTIMATION

The figure 3.9 depicts changes or errors in voltage magnitude where as it should be stable. At starting, there is some variation between 0.05 to 0.036 which gets enhanced, at 100 days, it goes upto 0.072 and then starts to decrease to 0.04 at 140 days. Therefore not useful parameter estimation in this period. Then from 180 days to 400 days, parameters find effective filtering and voltage magnitude becomes largely stable and is confined within the 0.03 to 0.04 range.

Figures 3.8 and 3.9 depicts noise and error or changes in the voltage magnitude. These parameters monitor the filters performances. In fig. 3.8 the noise in the extended Kalman filter reduces, but at starting, it is high. On analyzing fig.3.6 and fig.3.8, the reduction is approximately same as Kalman filter.

The figure 3.9 analyzes changes in voltage magnitude which should be stable. At starting it is not useful parameter estimation, but after that parameter estimation is effective and voltage magnitude is stable. But compared to fig 3.7, voltage magnitude stability is less as compared to the Kalman filter.

3.8 Performance of Unscented Kalman Filter

The UKF is simpler to calculate than the EKF [84] as there is no need to calculate the Jacobian matrix of the non-linear systems. Few years ago, the UKF had restricted applications in estimation of states in power systems. Recently, the unscented Kalman filter is implemented successfully in many power system applications and the efficiency of such filter is currently under review.

The Unscented Kalman filter is a highly effective discrete time based recursive filter which is used for solving the problems of estimation as below:

$$x_k = f(x_{k-1}) + q_{k-1} \quad \ldots \ldots \ldots \ldots \ldots \ldots \ldots \ldots \ldots \ldots (9)$$

$$y_k = h(x_k) + r_k \quad \ldots \ldots \ldots \ldots \ldots \ldots \ldots \ldots \ldots \ldots (10)$$

Where,

x = state vector and;

y = measurement vector.

h and f functions = non-linear equations of the measurement and system, respectively.

q_{k-1} = system noises with zero covariance and mean matrices Q and R.

r_k = measurement noises with zero covariance and mean matrices Q and R.

The UKF is a combination of the unscented transformation (UT) and the classical Kalman filter theory. The unscented Kalman filter (UKF) presents a more or less exact system rather than linearized models as in case of EKF which normally avoids the loss due to information based on higher order.

From the theory of the unscented transformation as described, the Sigma points $\{x_{k-u}\}$ are modelled and then evaluated step by step with the help of equations of the system

$$X_k = f(\chi_{k-1,i}) \quad i = 0, \ldots \ldots, n \quad \ldots \ldots \ldots \ldots \ldots (11)$$

Subsequently, we compute the state mean (predicted) vector x_k^-, and the covariance (predicted) matrix P_k^- as below:

$$x_k^- = \sum_{i=0}^{2n} W_i^m X_k^i \quad \ldots \ldots \ldots \ldots \ldots \ldots \ldots \ldots \ldots (12)$$

$$P_i^- = \sum_{i=0}^{2n} W_i^c \left[(X_k^i - x_k^-)(X_k^i - x_k^-)^T \right] + Q_{k-1} \ldots (13)$$

Where,

X_k^i = Matrix X_k based $(i+1)^{th}$ column

$X_k = n \times (2n+1)$ matrix which contains propagated form of sigma points.

One more step which involves the buildup concept of distinct sets of Sigma point $\{\chi_{k,j}\}$ that corresponds to P_k^- and x_k^- with same theoretical concept of UT the Sigma points are computed with the help of the given measurement equation:

KALMAN FILTER BASED PARAMETER ESTIMATION

$$Y_i = h(\chi_{k,j}) \quad \ldots \ldots \ldots (14)$$

The measurement covariance S_k matrix and the mean evaluated points μ_k and C_k represent the cross-covariance of the measurement and state and are calculated as follows:

$$\mu_k = \sum_{i=0}^{2n} W_i^m Y_k^i \quad \ldots \ldots \ldots (15)$$

$$S_k = \sum_{i=0}^{2n} W_i^c \left[(Y_k^i - \mu_k)(Y_k^i - \mu_k)^T \right] + R_k \quad \ldots \ldots (16)$$

$$C_k = \sum_{i=0}^{2n} W_i^c \left[(X_k^i - x_k^-)(Y_k^i - x_k^-)^T \right] \ldots \ldots \ldots (28)$$

Finally, the filter gain K_k, the updated forms of covariance P_k, and the state mean x_k are calculated as below:

$$K_k = C_k S_k^{-1} \quad \ldots \ldots \ldots (17)$$

$$x_k = x_k^- + K_k(y_k - \mu_k) \quad \ldots \ldots \ldots (18)$$

$$P_k = P_k^- - K_k S_k K_k^T \quad \ldots \ldots \ldots (19)$$

Unscented Kalman filter was applied to the IEEE 30 bus system, discussed above and the changes in the voltage level and noise was studied for different time frames.

Figure 3.10 depicts changes in noise. The parameter is monitored and results are based on filters performances for the time changes in days.

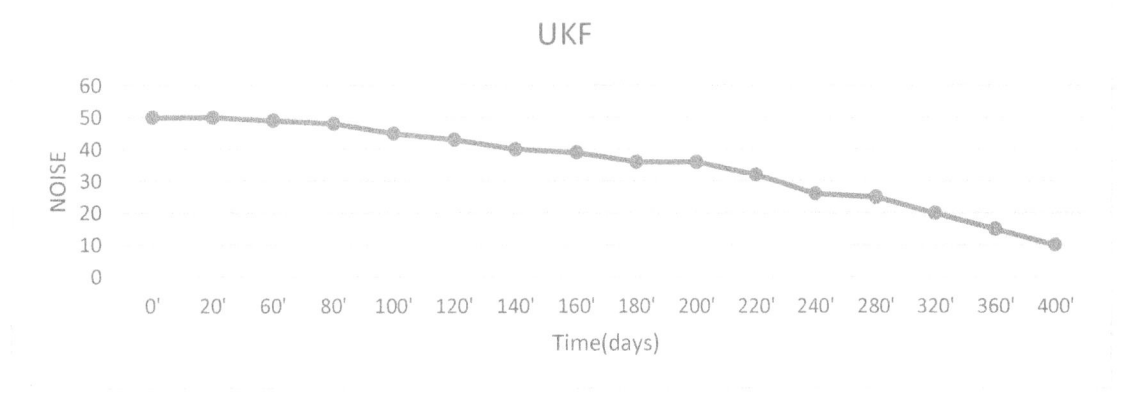

Figure 3.10 UKF response on noise during different time frames

KALMAN FILTER BASED PARAMETER ESTIMATION

In study of the Unscented Kalman filter in fig 3.10, it is seen that in the beginning, the noise is around 50 units per kunit of the signal but continuously decreases with the increase in time and at 400 days, settles down to around 10 units, which shows that noise reduces for Unscented Kalman filter as time increases, although this is not as effective as Kalman and extended Kalman filters.

Figure 3.11 depicts changes in the voltage magnitude. The parameter is monitored and results are based on filters performances for the time changes in days.

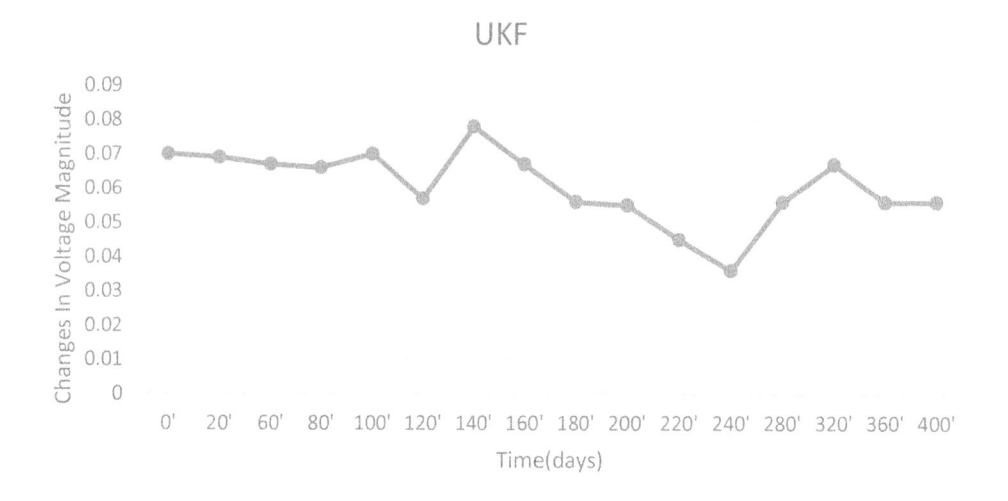

Figure 3.11 UKF response on Voltage during different time frames

The figure 3.11 analysis errors or changes in voltage magnitude which should be stable. At starting for first 100 days, the value is approximately 0.07 and then till 360 day, values are not stable, which means not effective parameter estimation but after that parameter estimation is found effective and voltage magnitude is stable from 360 to 400 days. But the filter depicts less stability as compared to Kalman filter.

3.9 Experimental Comparative Performance

Table 3.1 traces the noise dynamics with respect to the three approaches used mainly Kalman Filter, Extended Kalman Filter and Unscented Kalman Filter.

Table 3.1 Comparative Performance for Kalman Filter, Extended Kalman Filter and Unscented Kalman Filter with respect to noise

	Time (Days)	Kalman Filter	Extended Kalman	Unscented Kalman
Noise [per kunit]	20	44	44	50
	60	42	43	49
	100	32	37	45
	140	26	33	40
	180	18	24	37
	220	15	10	32
	260	10	15	26
	300	8	13	22

As observed in Table 3.1 the noise in Kalman filter speedily reduces from 44 units at 20 days to 8 units in 300 days. In EKF the reduction for the same period is from 44 units to 13 units, therefore EKF depicts comparable performance with respect to Kalman filter, but for UKF, the reduction is from an initial high of 50 units to 22 units which is again high as compared to noise in EKF and Kalman filter.

Table 3.2 traces the changes in voltage magnitude with respect to the three approaches used which is the Kalman Filter, Extended Kalman Filter and Unscented Kalman Filter.

Table 3.2 Comparative Performance for Kalman Filter, Extended Kalman Filter and Unscented Kalman Filter with respect to change in Voltage magnitude

	Time (Days)	Kalman Filter	Extended Kalman	Unscented Kalman
Error in the Voltage magnitude	20	0.024	0.042	0.07
	60	0.023	0.041	0.068
	100	0.025	0.038	0.07
	140	0.03	0.04	0.08
	180	0.024	0.032	0.058
	220	0.024	0.031	0.045
	260	0.024	0.032	0.045
	300	0.024	0.036	0.06

In Kalman filter voltage magnitude is stable as the error settled down to 0.024 . In EKF, besides the error being high shows higher variation from 0.042 to 0.036 as compared to Kalman filter. The UKF depicts still higher error levels from 0.07 to 0.06 and even going upto 0.08.Therefore UKF is still less stable compared to EKF and Kalman filter.

3.10 Conclusion

The Kalman Filter, dynamic state estimation method is able to provide an optimal solution to the process of real-time data prediction . The Phasor Measuring Units (PMU) play an important role in power transmission and distribution processes as they are able to monitor the power flow within a network. The PMU-based monitoring increases the efficiency of the smart grid. But there is a problem which is that the implementation of PMU enhances the dynamics of noise variance which further inflates the uncertainty in noise-based distribution. This chapter traces various extensions such as unscented and the extended forms of Kalman filter that specifically work on non-linear systems for dynamic state estimation.

In experiment shows the errors in voltage magnitude analysis in different period and also noise in PMU in different time periods. The analysis on comparison of performance of Kalman filter, Extended Kalman filter and Unscented Kalman filter with respect to noise, as per Table 3.1 depicts that out of the three, Kalman filter is most effective in reducing noise but comparable to EKF.

Also, in Table 3.2 ,the comparison of performance of Kalman filter, Extended Kalman filter and Unscented Kalman filter is done with respect to errors or change in Voltage magnitude and is seen that Kalman filter is able to reduce error most effectively.

Also to be noted that Kalman filter is a linear quadratic estimation technique whereas EKF and UKF are highly effective with non linear systems but the calculations are complicated, hence in the next chapter we work on linearizing Kalman filter estimation using Taylor series expansion and using the concept of Bayesian prediction.

Chapter 4   KALMAN FILTER ESTIMATION WITH BAYESIAN APPROACH

Kalman filter displayed better results compared to EKF and UKF for reducing the noise and errors in Voltage measurements. Kalman filter is a linear quadratic estimation technique whereas EKF and UKF are highly effective with non linear systems but the calculations are complicated, hence in this chapter we work on linearizing Kalman filter estimation using Taylor series expansion and using the concept of Bayesian prediction.

Kalman filter also qualifies for a very simple dynamic Bayesian network. We already know that Kalman filter is used to calculate the estimates of true values of states recursively for a period of time while using the incoming measurements and the mathematical process model. For Bayesian estimation, there is a recursive estimation of an unknown probability density Function using the incoming measurements and the mathematical process model.

4.1 Kalman Procedure

The nonlinearities of system-based dynamics are usually estimated in approx. by the linear version of non-linearized system model around the estimation of the last state. For a valid type of approximation, this process of linearization using the first order taylor series should be presented as a good approximation of non-linearized model in global uncertain domain linked with the estimation of state.

The Kalman filter (KF) theory is generally applied to the problem of linear-Gaussian, but most significant application of real world is non-Gaussian and/or nonlinear. Engineers generally use the linearized approximations in order to make such theory to get fit into the non-linearized issues, which are confronted in the applications of real world. Such kind of linearized approximations are very easy to be in use explaining the demand of EKF. An issue related to linear filter represents all the functions i.e. f, G, and h, which are linearized in x, and a "Gaussian" problem represents all the noise (v and w) are Gaussian in nature. The extended KF approximate (f, G, and h) the Taylor series (first-order) observed at the current (present) estimate of x. In reality, there is nothing like "the" extended Kalman Filter (EKF), but instead there is a lot of collection of EKFs in hundreds. Particularly, distinct EKFs can be modelled by using a large list of tricks-based on engineering such as different coordinate systems, different covariance matrix-based factorizations, higher order (or second order) corrections of Taylor series to the prediction of state vector and/or the update of measurement, state vector iteration updated by the use of measurements, distinct orders to use the sequential form of scalar value measurements in order to update the vector state, quasi-decoupling, tuning noise process and combinational joint effort of all, as well as the other whistles and bells discovered by analysts engineers in hope of enhancing the performance of EKF. In some cases, the practical form of tricks generally results in wide advancement, but in often cases there is no such type of improvement in result and they make the system worse in performance. The concept of distinct system coordinate for example, the Polar and the Cartesian may result in significantly distinct accuracy, that sometimes creates trouble for both the mathematicians and physicists, as the "physics" does not depend over what the system-based coordinate is used for describing it, and the experts of mathematics generally feel that the descriptions that are "coordinate free" are simple, more powerful and general. The non-linear and linear filter issues have the similar theoretical optimized accuracy of estimation in any of the coordinate system hence here the Kalman Filter is chosen for the Bayesean Approach.

4.2 IEEE 30 bus system

System considered for performance is the IEEE 30 Bus System

4.3 Performance of Kalman filter using Taylor series expansion

Taylor series represents a function as an infinite sum of terms which can be calculated from the function's derivatives values at a single point. The concept was developed by ta Scottish mathematician J. Gregory and was formally introduced in 1715 by Brook Taylor, an English mathematician. For the case where the Taylor series would canter at zero, then that series is referred to as a **Maclaurin series**. This was named after the Scottish mathematician Colin Maclaurin, who had worked extensively on this special case of Taylor series.

A function is normally approximated as a finite number of terms of its Taylor series. This kind of approximation gives rise to errors in quantitative estimates which is also given by **Taylor's theorem.** A polynomial is coined by taking few initial terms of the Taylor series and is denoted as the Taylor polynomial. The Taylor series of any function is the limit of that particular function's Taylor polynomials while the degree increase, provided the limit exists. It is not essential that the function be equal to the Taylor series, even though the Taylor series may converge at every point. The function which is equal to its Taylor series in called an open interval and is an analytic function in that interval.

Kalman filter is a Linear Quadratic Estimation technique and works on linear systems.

Here we have combined Kalman filter with Taylor series to linearize the process of dynamic state estimation.

Figure 4.1 depicts the change in noise for different number of measuring units for Kalman filter with Taylor expansion

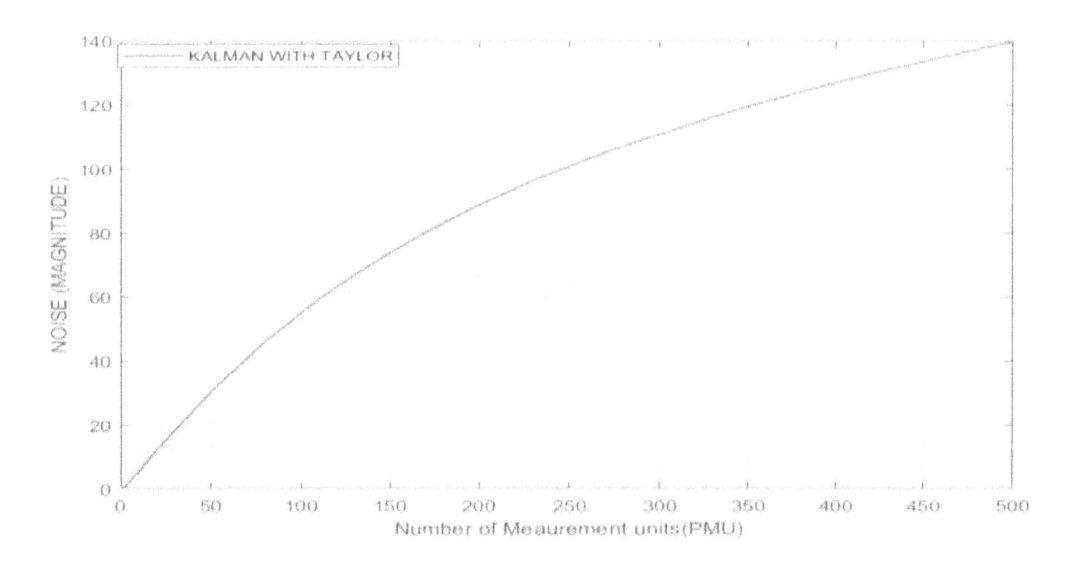

Figure 4.1 Kalman with Taylor response on Noise for different measurement units

As the number of measurement units increase, the noise increases. That is normally the case, as already stated that increase in the number of PMU's causes noise to increase. Taylor expansion tries to find efficient polynomials for reducing noise in PMU readings. Still, it is not able to optimize, so fistulation increases the sound when increasing the PMU in the bus system, and it also reduces the PMU utilization.

4.4 Performance of Kalman filter using Bayesian Approach

Bayesian approach is an approach which uses statistical inference where the popular Bayes'

Theorem is made use of to update the probability for a hypothesis continuously as more information and evidence becomes available. Bayesian prediction is used often in statistics, especially in mathematics statistics. Whenever there is a requirement of dynamic analysis of sequence of data, the Bayesian concept is utilized.

The Bayesian Probability, as it is called due to the subjective probability of Bayesian inference which finds application in a wide range of activities like engineering, philosophy, sports, law and medicine.

The Bayesian Probability calculates the posterior probability using two criteria, the first one is the prior probability and the second one is the likelihood function. The likelihood function is derived from the statistical model of the observed data.

Hyper parameter tuning is a kind of Bayesian optimization: minimize a function $f(\theta)$, by query values and compute gradients, Input θ: a configuration of hyper parameters, Function value $f(\theta)$: error on the validation set. Each evaluation is expensive. For a fixed adaptive buffer of evaluations, the optimizer performance is quantified by the improvement obtained after the optimization is completed that is when the evaluation buffer is fully consumed. Hence, for each iteration n, the best Bayesian optimization algorithm analyzes the leading buffer, at the end of the optimization process to the maximum possible improvement. To define this optimal Bayesian optimization algorithm, it is first of all essential to characterize, a design evaluated at iteration n which is capable of affecting the following iterations optimally.

Every buffer has a defined probability of occurrence which is characterized by a statistical model. Hence, for a known optimization policy, the improvement expected which is obtained at the end of the optimization can be quantified through simulation machinery. The optimal Bayesian optimization algorithm maps to a best optimization policy which provides solution for intractable dynamic programming.

Bayesian Filter

Bayesian filtering [11] models a problem as a dynamic system where the vector state $\mathbf{x}_n$, at discrete time n, represents the coordinates . At time n the a posteriori pdf Bel($\mathbf{x}_n$) of the state $\mathbf{x}_n$, called *belief*, is evaluated in two distinct steps. For the first step, here the belief function is updated as per the the *mobility model* $p(\mathbf{x}_n|\mathbf{x}_{n-1})$, this represents the dynamic model for the system and yields Bel($\mathbf{x}_n$). The mobility model describes the state variation $\mathbf{x}_{n-1} \to \mathbf{x}_n$, which is, the statistical description. For the second step, the belief function is again updated to Bel($\mathbf{x}_n$) which is to account for the statistical information $p(\mathbf{r}_n|\mathbf{x}_n)$ for the position at time n starting at the measurement vector $\mathbf{r}_n$ collected at time n. This is called the *perception model* and operates like an updater for the system state. Using the belief function, it is quite possible to identify the highly likely state (most stable position) at time n among all the possible states.

KALMAN FILTER ESTIMATION WITH BAYESIAN APPROACH

Recursive Bayesian state estimation (Bayesian filtering) [15, 48] is a known mathematical tool most often used in data fusion for performing tracking tasks. In this a general is described by the equations:

$$X_k = f_k(x_{k-1}, v_{k-1}) \ldots \ldots (9)$$

$$Z_k = g_k(x_k, \eta_k) \ldots \ldots (10)$$

where $x_k \in \mathbb{R}^n$ represents the state vector at step k; $z_k \in \mathbb{R}^m$ represents observations (or measurements); $v_k \in \mathbb{R}^n$ is an i.i.d. random noise with known (called process noise); and $\eta_k \in \mathbb{R}^m$ is the observation noise. The functions f_k and g_k are, respectively, the process and observation models; usually they are considered time invariant. Thus, $f_k = f$ and $g_k = g$. To estimate the $p(x_k, z_{1:k})$, Bayesian filtering operates basically in two steps: prediction and update. The system transition model $p(x_k|x_{k-1})$ and the set of available observations $z_{1:k-1} = \{z_1, \ldots, z_{k-1}\}$ provide the posterior prediction as

$$p(x_k|z_{1:k-1}) = \int p(x_k|x_{k-1}) p(x_{k-1}|z_{1:k-1}) dx_{k-1} \ldots \ldots (11)$$

New observations z_k at time k and the observation model supply the likelihood probability $p(z_k|x_k)$, which is used to correct the prediction by means of the update process:

$$p(x_k, z_{1:k}) = p(z_k|x_k) p(x_k|z_{1:k-1}) p(z_k|z_{1:k-1}) \ldots \ldots (12)$$

It is a known fact that a closed-form optimal solution to this problem is achieved using the hypothesis of linearity of the state, an observation model, and the measurement and process noises. But, these conditions occur rarely in real-world situations, the Kalman filter is hence successfully employed in many applications wherever the target movements are reasonably smooth. Also, to cope with some more general situations many algorithms are proposed in literature which give an approximate solution.

KALMAN FILTER ESTIMATION WITH BAYESIAN APPROACH

Bayesian Pseudocode

Algorithm 1: Bayesian Approach
Input: Deploy BUS With Grid
Output: Optimize queue length for avoiding collision.

1). Deploy Bus with grid

2). Deploy Load, BUS and variation.

3). In every BUS there are multiple Lines and every Lines has noise n_s and total noise N_C.

4). Initialize noise priority according to number of Lines. If packets are same then choose randomly.

5). From step 4 Initialize Bayesian approach

$$P(q_p / I_p) = \frac{P(q_p) * P(I_p / q_p)}{P(I_p)} \quad \ldots\ldots\ldots\ldots 1$$

$P(q_p)$ = new priority of queue.

$P(I_p)$ = Initialize priority of noise.

6). Step 5 Start giving prediction of priority $P(q_p)$ and q_p.

6.1 $$P(q_p / P_i) = \frac{(P_i) * P(P_i / Q_P)}{\sum P(P_i)} \quad \ldots\ldots\ldots\ldots 2$$

$P(q_p / P_i)$ = iterative noise priority on Packet P_i

6.2 $$q_p = \frac{\sum P(q_p / P_i)}{n}$$

7). Analyse noise and voltage.

KALMAN FILTER ESTIMATION WITH BAYESIAN APPROACH

In figure 4.2 depicts the noise performance when increasing the PMU's using Kalman with Bayesian approach

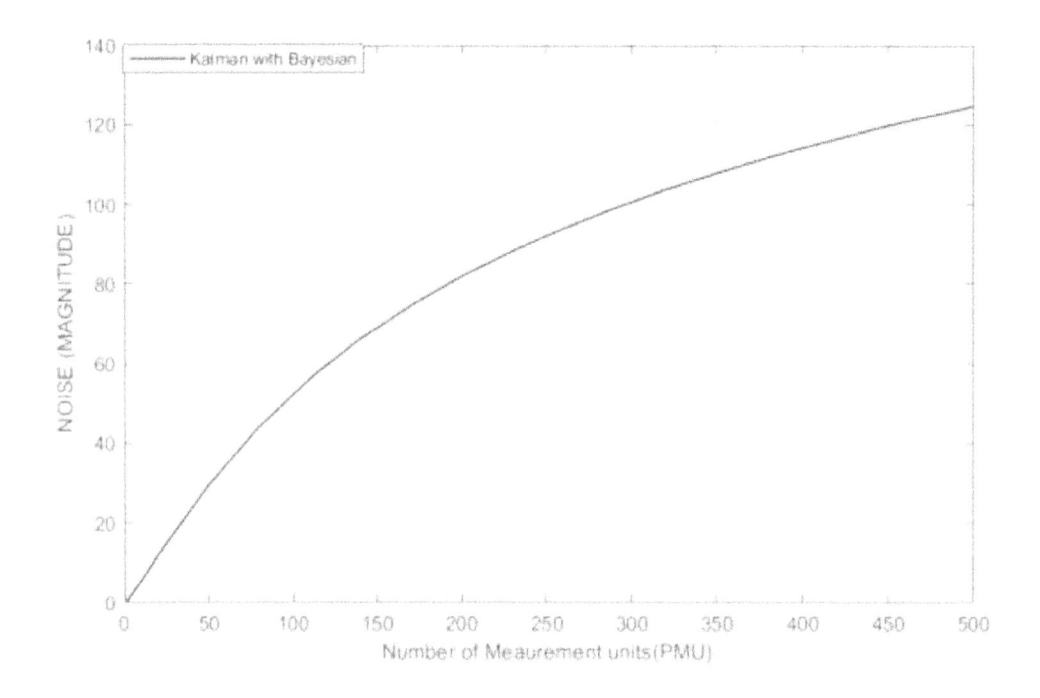

Figure 4.2 Kalman with Bayesian response on Noise during different measurement units

Bayesian approach finds dependency between efficient polynomials and PMU values. It reduces noise in PMU readings but it will not optimize, but increases the noise when increasing the PMU in the bus system, hence reduces the PMU utilization, but compared to Taylor's expansion, this approach improves noise or reduces noise on increasing the PMU.

4.5 Performance of Kalman filter using Bayesian with Taylor expansion

Bayesian approach finds the dependency between efficient polynomials and PMU values and Taylor expansion finds the sufficient polynomial value of the Kalman filter.

The figure 4.3 depicts the noise performance when increasing the PMU's using Kalman with the Bayesian approach with Taylor expansion.

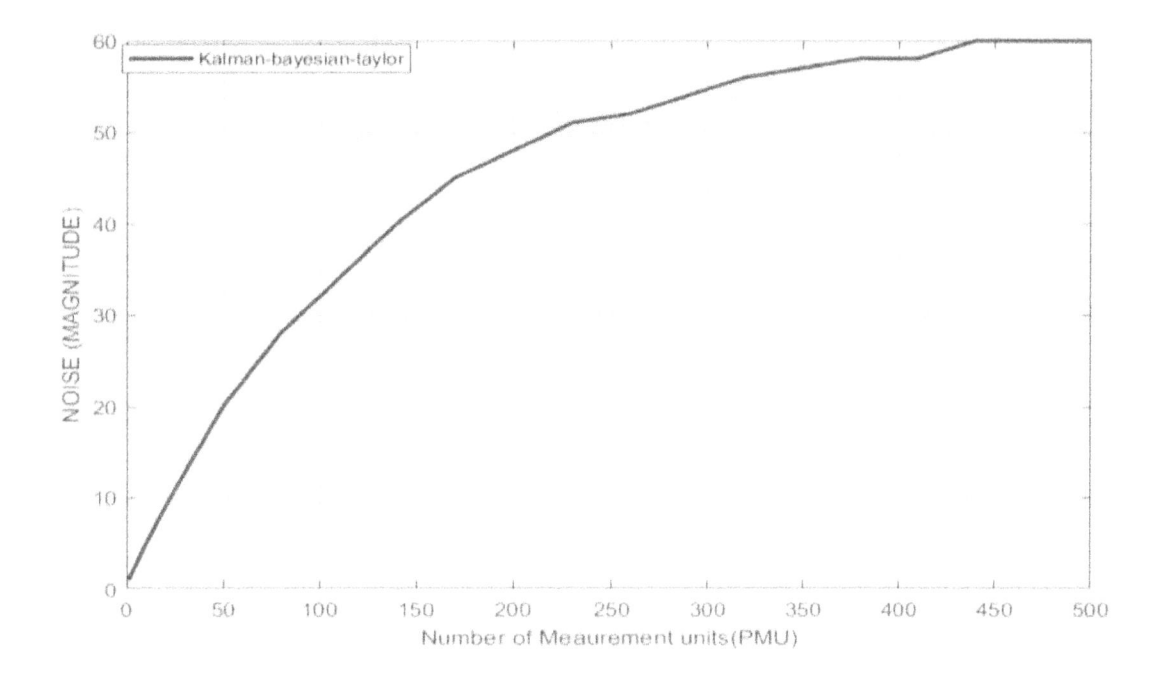

Figure 4.3 Kalman-Bayesian-Taylor response on Noise during different

Bayesian approach finds the dependency between efficient polynomials and PMU values and Taylor expansion finds the sufficient polynomial value of Kalman filter. Once again, it increases the noise when increasing the PMU in the bus system, and it also reduces the PMU utilization, but compared to the Taylor's expansion, this approach improves noise or reduces noise when increasing the PMU, but does not really optimize the readings.

4.6 Comparative Performance

The figure 4.4 depicts the comparison of the performance of Kalman -Taylor, Kalman-Bayesian and Kalman-Bayesian -Taylor approach on reducing noise when increasing the number of PMU's

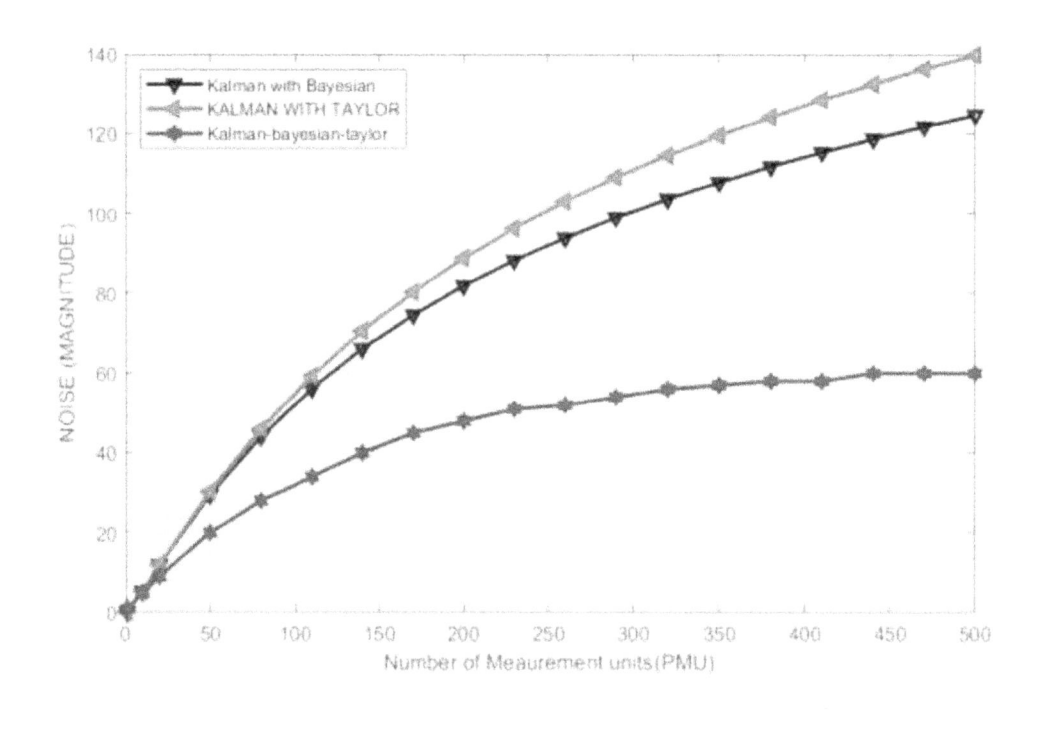

Figure 4.4 Comparison of filters response on Noise for different number of measurement units

As seen in the graph, right at the beginning the Kalman-Taylor and Kalman-Bayesian graphs show the same results, almost overlapping till 80 PMU's but in this period Kalman -Bayesian-Taylor shows lesser noise, like at 20 PMU the noise for Kalman -Bayesian-Taylor is 20 units per kunit of signal where as for Kalman -Bayesian it is 28 units and Kalman-Taylor it is 30 units. Subsequently as number of PMU increase this gap widens, at 200 PMU for Kalman -Bayesian-Taylor it is 48 units, for Kalman-Bayesian it is 85 and Kalman-Taylor it is 90.This increase continues where Kalman-Taylor displays maximum noise followed by

Kalman-Bayesian and for Kalman -Bayesian-Taylor, not only the noise is minimum but also becomes constant after 350 PMU which is not the case with the others.

Bayesian approach finds the dependency between efficient polynomials and PMU values and Taylor expansion finds the sufficient polynomial value of Kalman filter. Hence the Kalman -Bayesian-Taylor approach improves noise or reduces noise when increasing the PMU. Other methods like Kalman with Bayesian and Kalman with Taylor do not improve the noise as much because they are not able to optimize or find the relation between PMU and noise.

The figure 4.5 depicts the comparison of the performance of Kalman -Taylor, Kalman-Bayesian and Kalman-Bayesian -Taylor approach for reducing Mean Square Error when increasing the number of PMU's .

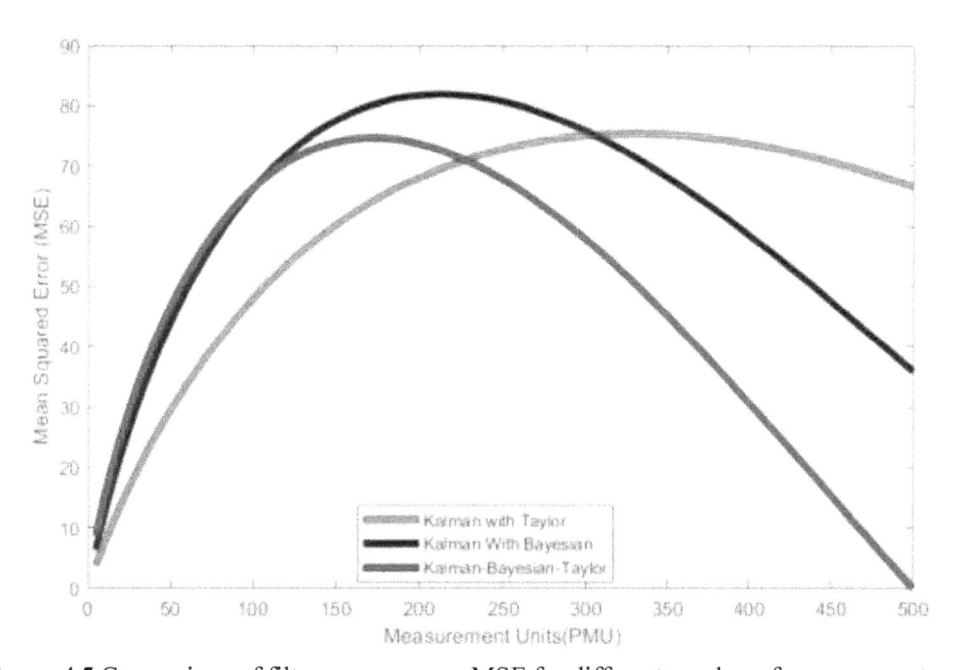

Figure 4.5 Comparison of filters response on MSE for different number of measurement units

From the graph it is observed that in the beginning the Kalman-Bayesian-Taylor and Kalman-Bayesian graphs overlap, they depict the same Mean square error till about 120 PMU where as the MSE of Kalman-Taylor is much less. After 120 PMU, the increase with Kalman-Bayesian is higher than Kalman-Bayesian-Taylor but after 170 PMU the MSE for both starts

KALMAN FILTER ESTIMATION WITH BAYESIAN APPROACH

to decrease and this is more pronounced for Kalman-Bayesian-Taylor. The MSE for Kalman Taylor continuously increases and at 230 PMU it rises above Kalman-Bayesian-Taylor and at 330PMU it rises above Kalman-Taylor too.

Bayesian approach finds the dependency between efficient polynomials and PMU values and Taylor expansion finds the practical polynomial value of Kalman filter. In reducing noise in PMU readings it does not optimize, so increases the mean squared error (MSE) when increasing the PMU in the bus system, and it also reduces the PMU utilization, but later in the system the mean square error decreases in the system due to optimization of Bayesian convergence and gets effective threshold of polynomial value. Other methods like Kalman with Bayesian and Kalman with Taylor do not improve MSE as much because of not being able to optimize and find the relation between PMU and MSE.

Table 4.1 Comparison of Kalman Taylor, Kalman Bayesian and Kalman Bayesian Taylor

	No. of PMU	Kalman Taylor	Kalman Bayesian	Kalman Bayesian Taylor
Noise [per kunit]	100	52	50	32
	200	90	85	48
	300	112	84	54
	400	128	112	58
Mean Square Error [per kunit]	100	47	67	67
	200	68	84	74
	300	77	77	58
	400	75	60	32

4.7 Conclusion

Phasor measurement method is highly significant for achieving an efficient performance with respect to the smart grid technology. But despite these sophisticated measurement methods, there are inconsistencies experienced during measuring operations as the paramaters measured show divergent results. Phasor Measurement Unit (PMU) plays an important role in smart grid technology, where it contributes to measure the synchro phasors thus making it possible to dynamically monitor different types of transients occurring in a system. Here we have compared the popular Kalman filter with Taylor expansion technique with a novel method of Kalman filter Covariance Bayesian learning and also Kalman with Bayesian with Taylor expansion . The noise increases with the number of PMU's in all the three cases as is said that increasing the number of PMU's though makes the system more observable but at the same time increases the noise dynamics. In the three cases compared above the noise increases with the number of PMU's but the increase is least in the case of Kalman Bayesian Taylor approach and maximum with Kalman Taylor. Bayesian approach finds the dependency between efficient polynomials and PMU values and Taylor expansion finds the effective polynomial value of Kalman filter. It reduces noise in PMU readings but it will not optimize so increases the mean square error (MSE) when increasing the PMU in bus system. Kalman Bayesian Taylor compared to Kalman Taylor expansion and Kalman Bayesian reduces noise on increasing the PMU and also reduces the MSE. Initially MSE increases and then decreases because at starting optimization between Kalman and Bayesian not reached but at convergence point after some time when the effective threshold is reached noise and MSE will decrease. Other approaches like Kalman with Bayesian and Kalman with Taylor are not able to improve MSE because of their inability to optimize on finding the relation between MSE and PMU by themselves.

Chapter 5   OPTIMIZATION OF KALMAN FILTER USING PARTICLE SWARM OPTIMIZATION

Kalman-Bayesian-Taylor was effective in reducing the noise and MSE to a certain extent compared to Kalman-Taylor and Kalman-Bayesian. To achieve further optimization, another approach of using an optimization technique where the estimation of the dynamic polynomial prior is done by using Particle Swarm Optimization (PSO) is explored for the Kalman filter.

5.1 Optimization Techniques

In the analysis of Integer Quadratic Programming (IQP), the problem of OPP is usually described by using the matrix-based connectivity that represents the topology of the network. The optimization of quadratic objective function that considers the linearized constrained along with consideration of integer-based variable value. This type of process will help in reduction of PMUs number by providing a full observability of the system under normal operating conditions and the outage condition.

Integer Linear Programming techniques, usually called as Binary integer programming, considers both the power calculation and injection of the system additionally with failure error calculation of PMU with the help of state estimation (SE). The process of ILP formulation is usually based on the values of eigenvectors usually obtained from the adjacency matrix of the spanning tree. In order to curtail the rate of installation, after the process of decomposition, using the ILP theory, the PMUs optimally reduce noise and MSE parameters in the sub networks of the system.

Greedy Algorithm: An algorithm based on combinable optimization that considers an appropriate form of local, or immediate, possible solution while searching the answer, is known as GA i.e. greedy algorithm. An algorithm based on matrix reduction and a virtual data pre-processing elimination approach has been introduced for decreasing the computing effort and placement model size for determining the set of optimized noise reduction.

5.2 Particle Swarm Optimization

PSO is a developmental computational method initially created by Eberhart and Kennedy (1995). The PSO is the stimulation of social conduct rather than progression of nature as in the other evolutionary algorithms (evolutionary programming, genetic algorithms, genetic programming, and evolutionary strategies). PSO is inspired sociologically, since the algorithm depends on sociological conduct related with the mechanism of bird flocking. It is a populace based evolutionary algorithm. Like the other populace based evolutionary algorithm, PSO is introduced with a population of arbitrary arrangements.

Advantages of PSO:

i) It is utilized for discrete/persistent or discontinuous continuity or convexity. It is less delicate to the idea of the objective function, for example convexity or continuity.

ii) It represents a derivative free algorithm not at all like numerous ordinary procedures.

iii) It has less parameter to alter not at all like numerous other evolutionary procedures. There are no mutation and crossover administrators as in GA and evolutionary-based programming.

iv) The evaluation of function is by methods by logic operations and basic mathematic. It is anything but difficult to actualize.

v) It has the adaptability to be coordinated with other techniques of optimization to frame a hybrid device.

vi) It involves the potential of effective computation with very a huge number of simultaneously working processors. So, it gets the capability to escape localized minima.

vii) It can grip the objective functions with stochastic conduct.

viii) It does not need a good preliminary key to assure the convergence procedure. [89]

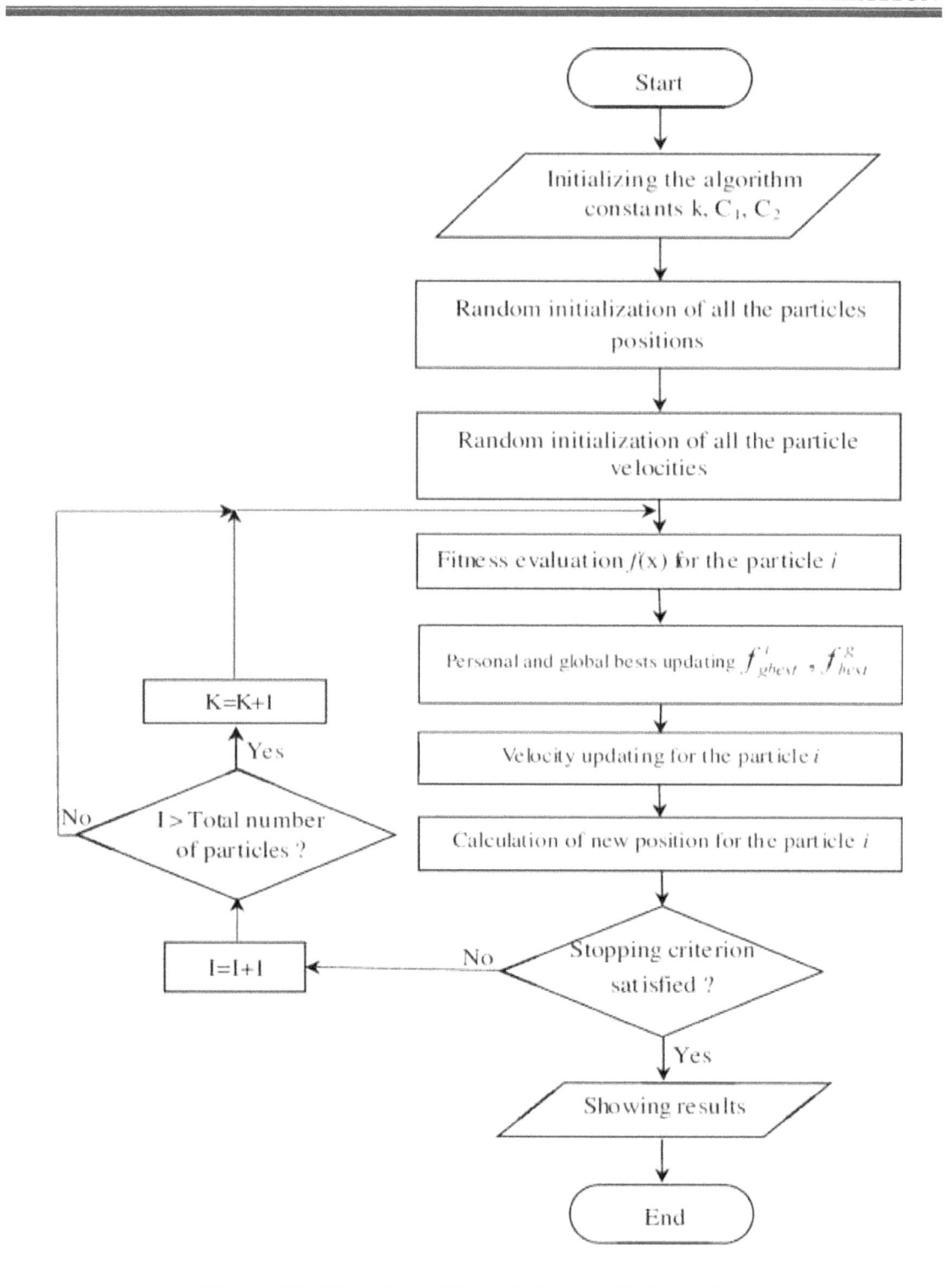

Figure 5.1: Flowchart of Particle Swarm Optimization

5.4 Kalman -Bayesian -Taylor PSO

- Kalman filter is a method which consists of continuous steps of prediction and filtering. The steps are interpreted and derived dynamically while using a framework based on Gaussian probability density functions. As per the system conditions, the Kalman filter converges towards a steady state filter and achieves steady-state. This innovative filtering process displays a novel information which is used for the state estimate along with the last measurement of the system Thus the filter dynamics can be interpreted with respect to the error ellipsoids related to the Gaussian pdf . But if the system state dynamics is nonlinear, the conditional probability density functions which provide the minimum mean-square estimate will no longer be Gaussian. The optimal non-linear filter will then propagate these non Gaussian functions and calculate their mean, but this leads to a high computational burden.

- The Bayesian technique normally treats this as a random function and places a prior over it. The prior is used to capture beliefs regarding the behavior of the function. After evaluating the function, which is now treated as data, the prior is then updated to form a posterior distribution for the objective function. Now the posterior distribution can be used to construct an acquisition function which will determine the next query point.

- There can be a number of methods used for defining the prior or posterior distribution over an objective function. A popular method used is a method called Kriging. There is another less expensive method which uses the Parzen-Tree Estimator for construct two distributions for the 'high' and 'low' points and subsequently finds a location which maximizes the expected improvement.

- Functions are approximated using a finite number of terms in its Taylor series. Taylor's theorem is used for quantitative estimates of errors introduced due to such an approximation. The Taylor polynomial is formed by taking some initial terms of the Taylor series. The Taylor series of a function is represented by the limit of a function where the Taylor polynomials in terms of degree increases

- considering the limit exists. Even though the Taylor series may converge at every point, a function may still not be equal to its Taylor series.

- Below given equation (1) represents the Kalman polynomial value before optimization and Bayesian prediction. X_k represent the predicted Bayesian value. A_K and A_{K-1} represent Kalman values at the current iteration and previous iteration. G_K^T show the predicted polynomial value at K iteration and T time. U_K initial noise predicted by Kalman filter. In equation (2), P_i represents the predicted value of swarm optimization. X_k is an initial value which predicts by equation (1).

- Equation (4) finds the location of a particle by a random function g_d and predicted value $X_{i,d}$.

- Equation (5) finds the optimal position represented by $V_{i,d}$. This optimal position defines the polynomial value of Kalman. ω represents the initial weights that come by the noise-based initial value of PMU. $\emptyset_p$ and r_p represents the initial parameter of swarm intelligence, these values signify local optimization. Its value generally varies between 0 and 1. $P_{i,d} - X_{i,d}$ expression finds the local and global optimization of the polynomial predicted value. $\emptyset_g$ and r_g signifies the initial parameter of swarm intelligence. These values characterize the global optimization. In equation (6), its Bayesian initialization calculates after prediction the optimized Kalman polynomial value. $f(x_{i|i-1}, \mu_k)$ expression find the relation between $x_{i|i-1}$ polynomial value μ_k and noise.

- Optimal Kalman

  - $X_k = A_K | A_{K-1} G_K^T U_K$ .... (1)
  - $P_i \leftarrow X_K \{X_K \text{ is initial Position of Particle}\}$ .... (2)

- Initialize the velocity

- $g \leftarrow P_i$ .... (3)

- g is random function

- Update particle Velocity

- $(g_d - X_{i,d})$ .... (4)

- $V_{i,d} = \omega V_{i,d} + \emptyset_p r_p(P_{i,d} - X_{i,d}) + \emptyset_g r_g$ ....(5)

When existing the optimization then following equation optimized

- $X_i| X_{i-1} = f(x_{i|i-1}, \mu_k)$ ....(6) From equation (5)

5.4 IEEE 30 Bus System

The IEEE 30 bus system is considered, same specifications as earlier

5.5 Performance of Kalman with PSO

Fig 5.2 depicts the noise magnitude changes for Kalman PSO. These experiments utilizes two different types of time parameters, the small-time unit, is described in figure 5.2. It presents a whole day examination with effective milestone analysis in noise magnitude

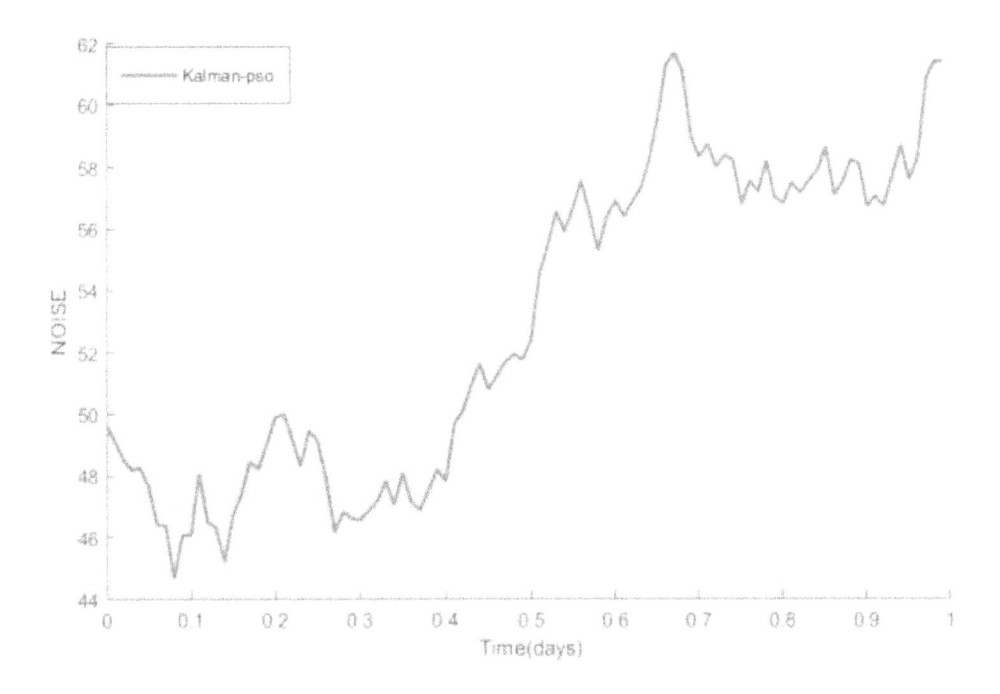

Figure 5.2 Kalman-PSO noise in a small-time frame

Figure 5.2 depicts the hybridization of the Kalman filter with PSO to optimize the polynomial value of the Kalman filter. Analysis results in an increase in variation of noise magnitude. In the beginning the noise magnitude is less ranging from 45 units to 50 units for 0.4 part of the day, but later it continues to increase to 62 units at 0.7 part of the day. After this there is a decline and becomes relatively stable around 58 units till 0.95 part of the day.

This experiment uses two different types of time parameters where the second is a large time unit, as depicted in figure 5.3 It represents an analysis of 500 days.

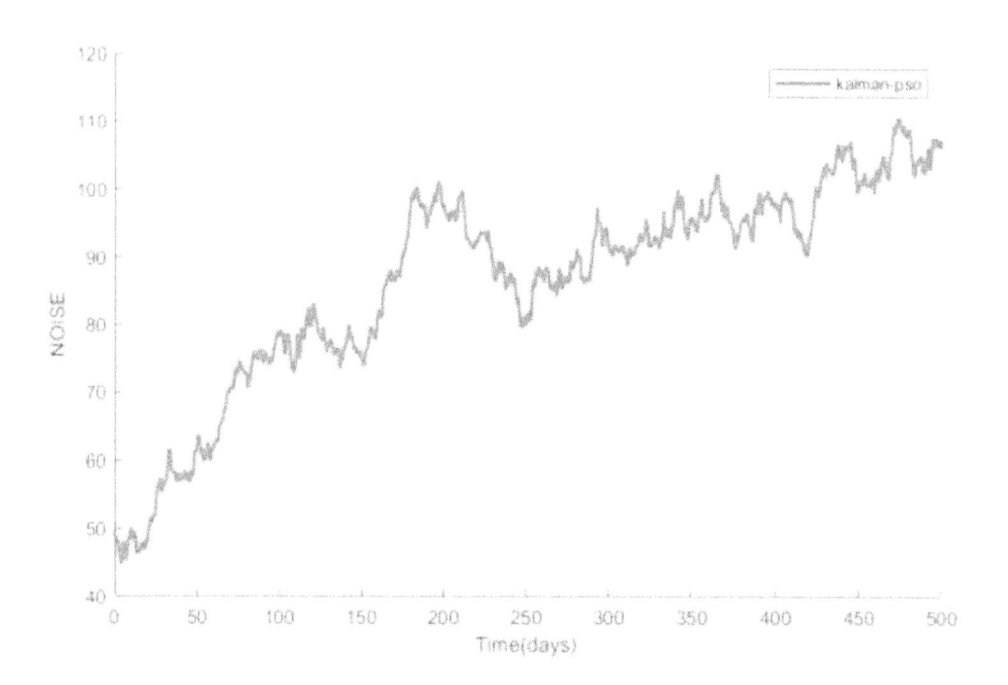

Figure 5.3 Kalman-PSO noise in a large time frame

In figure 5.3, the Kalman filter is optimized by PSO which takes the decision of polynomial value of the Kalman filter. Analysis results in an increase in variation of noise magnitude and an increase in noise from 50 units to 105 units as the time increases.

5.6 Performance of Kalman -Taylor PSO

Figure 5.4 depicts the changes in noise magnitude for the Kalman Taylor PSO approach. These experiments utilize two different types of time parameters where one is a small-time unit, as described in figure 5.4 It presents a whole day examination with effective milestone analysis in noise magnitude.

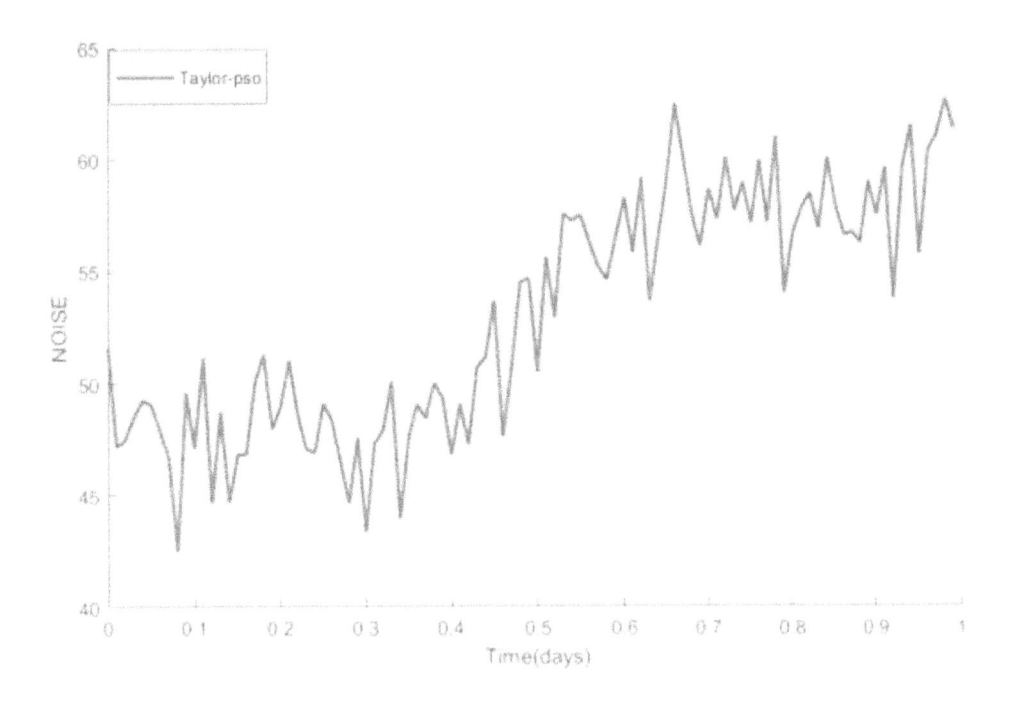

Figure 5.4 Kalman-Taylor-PSO noise in the small-time frame

Figure 5.4 depicts the hybridization of the Kalman filter with Taylor expansion and it helps to optimize the polynomial value of the Kalman filter. Analysis results in a constant increase in variation of noise magnitude in the range 42 units to 62 units throughout the time period of a day.

Figure 5.5 depicts the noise magnitude changes for a larger time unit. It represents an analysis of 500 days. It presents a whole day examination with effective milestone analysis in noise magnitude.

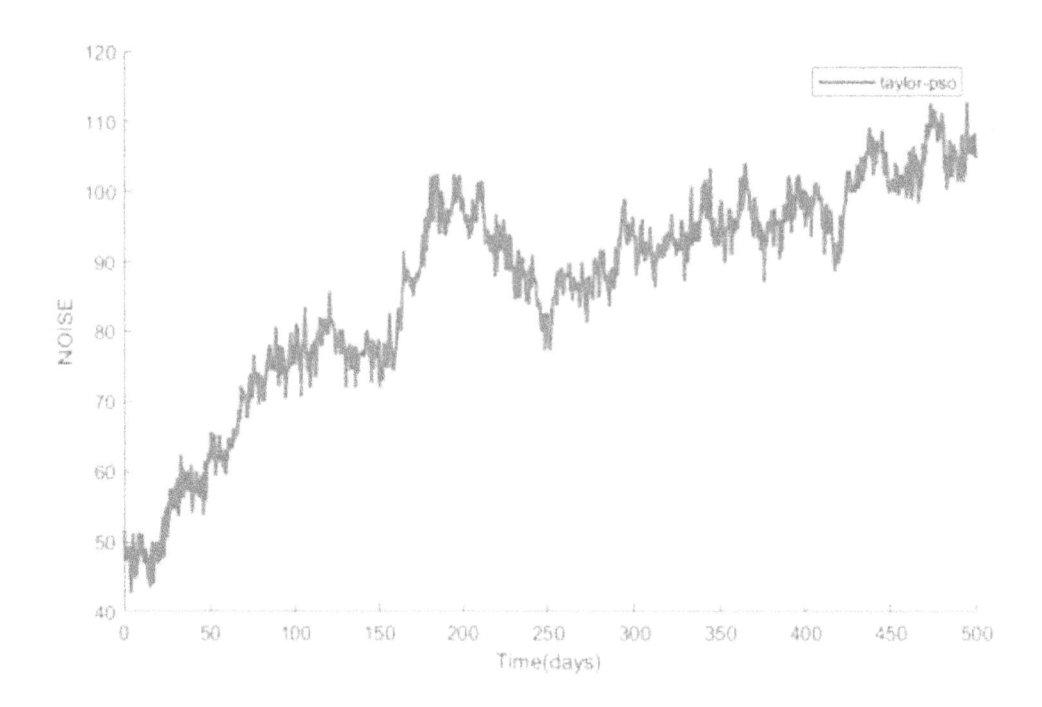

Figure 5.5 Kalman-Taylor-PSO noise in a large time frame

In fig 5.5, the Kalman Taylor PSO approach is used which helps to optimize the polynomial value of the Kalman filter. Analysis results in a constant increase in variation of noise magnitude which again varies between 42 units to 112 units.

5.7 Performance of Kalman -Bayesian-Taylor PSO

Figures 5.6 and 5.7 depicts the changes in noise magnitude for the Kalman-Bayesian-Taylor PSO approach. Analysis is for the two different types of time parameters where one is a small-time unit, as described in figure 5.6.

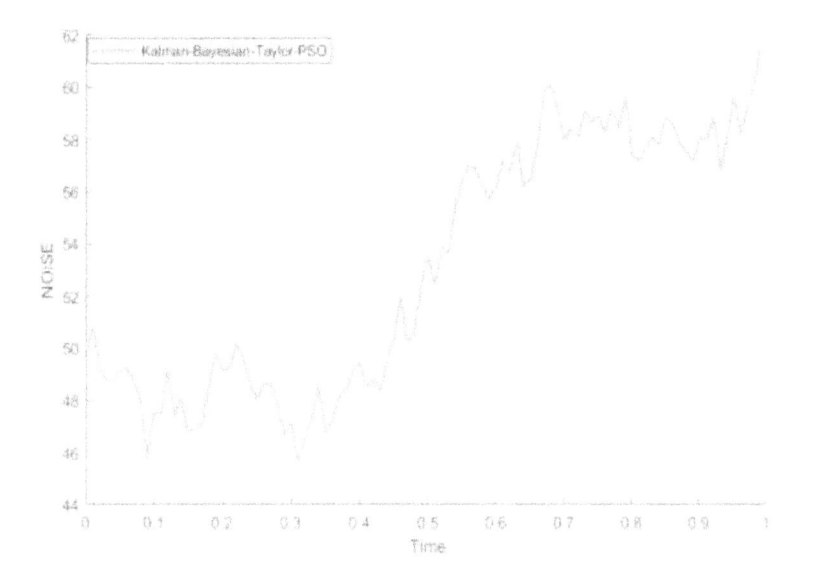

Figure 5.6 Kalman-Bayesian-Taylor PSO noise in a small-time frame

Fig 5.6 depicts the hybridization of the Kalman filter with Bayesian and Taylor approach and it helps to optimize the polynomial value of the Kalman filter. Here too analysis results in an increase in variation of noise magnitude which ranges between 46 units to 61 units. The graph depicts a gradual increase with some high and low in the beginning.

OPTIMIZATION OF KALMAN FILTERUSING PARTICLE SWARM OPTIMIZATION

Figure 5.7 depicts the noise magnitude changes for the Kalman-Bayesian-Taylor PSO for a large time frame. It represents an analysis of 500 days.

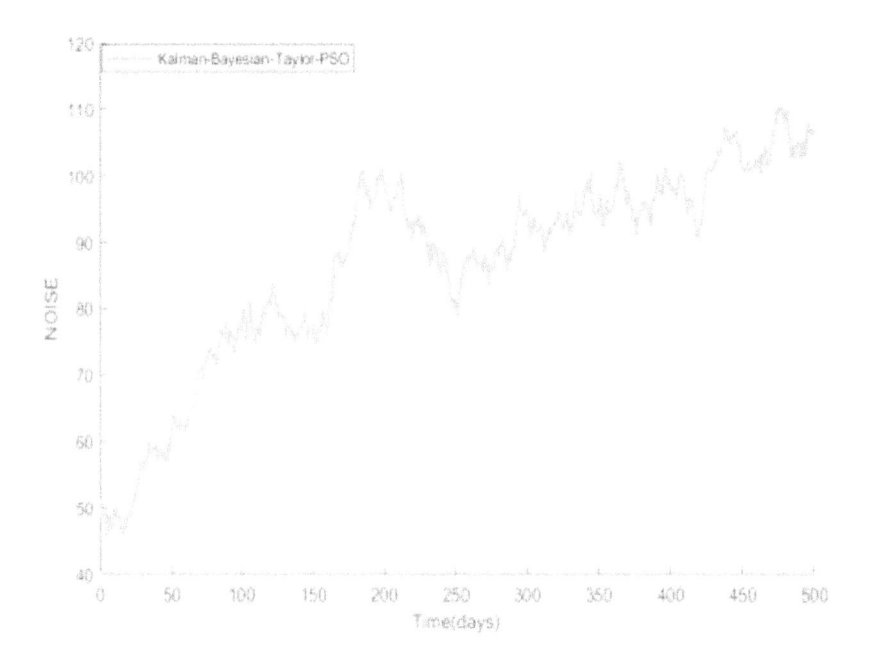

Figure 5.7 Kalman-Bayesian-Taylor PSO noise in a large time frame

Figure 5.7 signifies the hybridization of the Kalman filter with Bayesian and Taylor approach and it helps to optimize the polynomial value of the Kalman filter. Analysis results in a gradual increase in variation of noise magnitude from 48 units to 110 units for the days considered. It is cleared day-by-day and the proposed hybrid approach thereby improves the noise slightly as compared to other hybrid approaches.

5.7 Comparison of Hybrid Approach

In this section, a comparison of different hybrid approaches using the Taylor and Bayesian approach is done with the Kalman filter. In hybrid strategies, the use of particle swarm optimization is done by default. Using this experiment, analyses are done by particle swarm optimization (PSO) finding sufficient polynomial value in three of the following cases:

- Only with Kalman
- Kalman with Taylor
- Kalman with Bayesian and Taylor

Figure 5.8 shows the comparison of the above three approaches for small time. Analysis results in an increase in noise as the number of days increase.

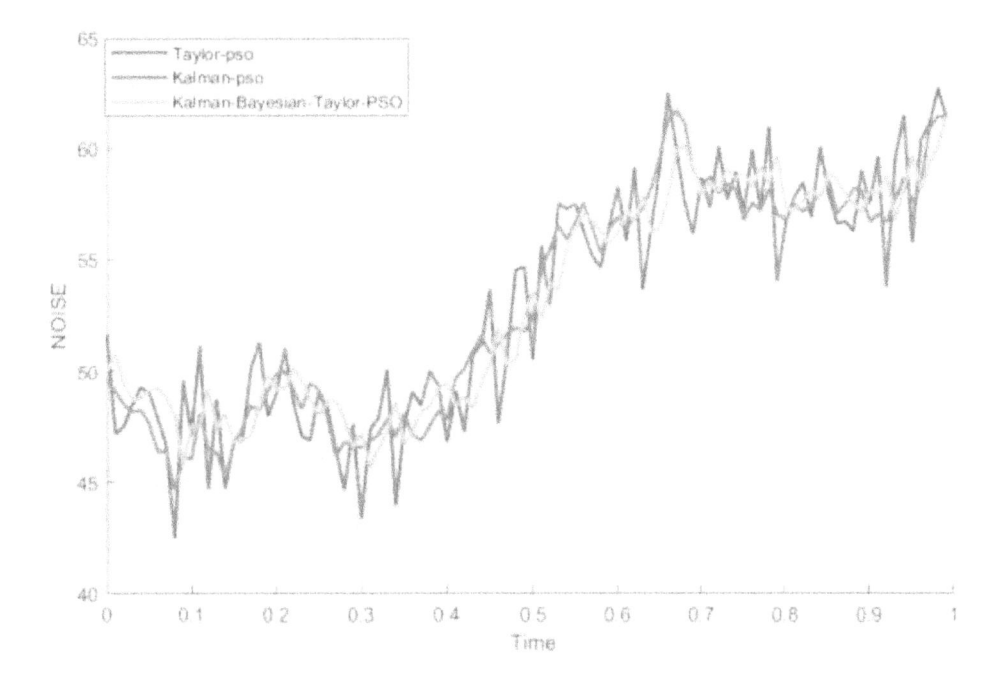

Figure 5.8 Comparison of Hybrid Kalman-filter noise in the small-time frame

The Taylor expansion approach improves the Jacobian matrix and enhances the gradient-based optimization. On comparing the hybridization of Kalman-Bayesian-Taylor with PSO, it is possible to find the effects related Bayesian approach and reduce the noise rapidly. Here, it is perceived that the Kalman-Bayesian-Taylor PSO is able to reduce the noise variations somewhat as compared to Kalman-Bayesian PSO and Kalman-Taylor PSO.

Table 4.1 Comparative Performance of Hybrid Kalman filter in the small-time frame

	Time (Day)	Kalman PSO	Kalman-Taylor PSO	Kalman-Bayesian-Taylor PSO
Noise (/Kunit)	0.1	46	48	45
	0.2	50	48	48
	0.3	47	44	47
	0.4	49	47	49
	0.5	52	55	51
	0.6	57	55	54
	0.7	59	58	57
	0.8	57	61	57
	0.9	57	58	57

The Table 4.1 depicts the comparative performance of the Kalman-PSO, Kalman-Taylor PSO and the Kalman-Bayesian-Taylor PSO in the small time frame. It shows that the Kalman-Bayesian-Taylor PSO is able to reduce the noise variations somewhat as compared to Kalman-Bayesian PSO and Kalman-Taylor PSO.

Figure 5.9 shows the comparison of the above given three approaches for larger time frame of 500 days.

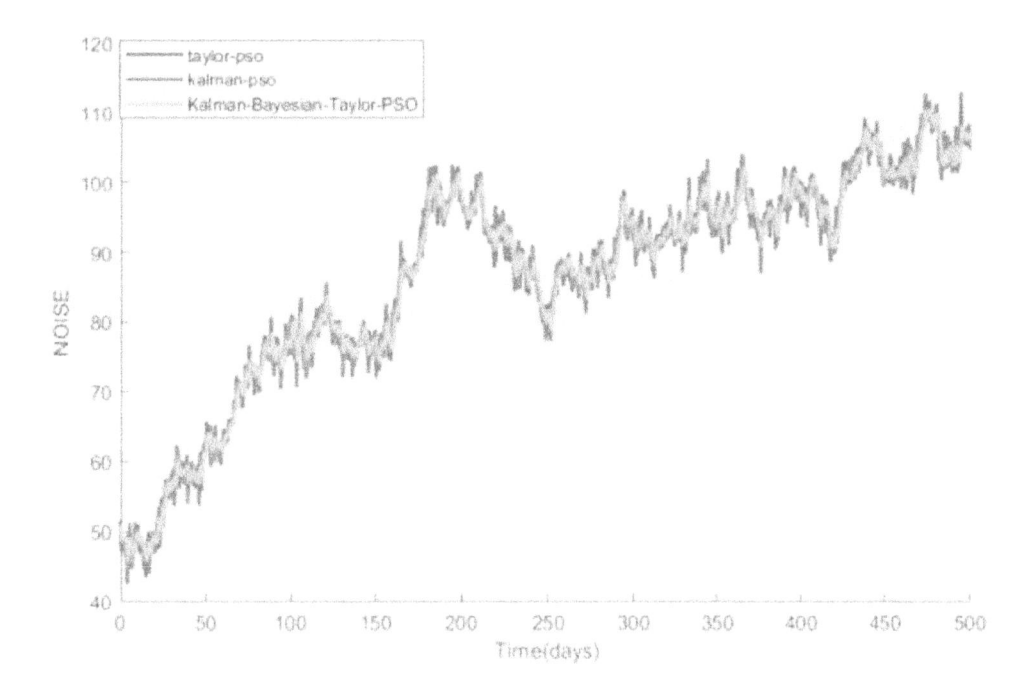

Figure 5.9 Comparison of Hybrid Kalman-filter noise in the large time frame

The Taylor expansion approach improves the Jacobian matrix and enhances the gradient-based optimization of the Kalman approach. On comparing the hybridization of Kalman-Bayesian-Taylor with PSO, it is possible to find the effects related Bayesian approach and reduce the noise compared to other two approaches.

Table 4.2 Table for Comparative Performance of Hybrid Kalman filter in the large-time

	Time (Days)	Kalman PSO	Kalman-Taylor PSO	Kalman-Bayesian-Taylor PSO
Noise (/Kunit)	50	58	59	58
	100	75	75	74
	150	80	78	78
	200	95	95	94
	250	80	88	78
	300	88	98	86
	350	95	95	92
	400	99	98	96
	450	102	98	98

The Table 4.2 depicts the comparative performance of the Kalman-PSO, Kalman-Taylor PSO and the Kalman-Bayesian-Taylor PSO in the large time frame. The Kalman-Bayesian-Taylor PSO is able to reduce the noise variations somewhat as compared to Kalman-Bayesian PSO and Kalman-Taylor PSO.

5.8 Conclusion

In this chapter, the analysis of the proposed hybrid and hybrid optimization approach was made by using particle swarm optimization. PSO finds the effective polynomiality of the filter for reducing noise parameters. The experiment analyzes different time-based noise analysis. In figure 5.8 and figure 5.9, comparative analysis of short time and long period noise variation is depicted using hybrid strategies with PSO. The analysis has been tabulated in Tables 4.1 and 4.2, where we see that the Kalman-Bayesian-Taylor PSO is able to reduce the noise variations somewhat as compared to Kalman-Bayesian PSO and Kalman-Taylor PSO.

Chapter 6   EXTENDED KALMAN FILTER

In this chapter, development of non-linear extensions of Kalman Filter like the Extended Kalman filter is explored, which can handle the non-linearity aspect of the problem for dynamic state estimation.

6.1 Introduction

In the estimation hypothesis, EKF i.e. Extended Kalman Filter represents a nonlinear adaptation of KF which helps to linearize the current covariance and mean estimation. On account of the characterized transition models, EKF is viewed as the true standard in the hypothesis of navigation frameworks, GPS enabled EKF channel, and nonlinear state estimation is presented to take care of the issue of non-linearity in KF. All things considered, there might be a lot of situations where the framework may glance in one area and it may take the estimation from another perspective. This includes angles or edges to take care of the issues, bringing about the non-linear function which when provided to the Gaussian results in a non-Gaussian type distribution. Also, over a non-Gaussian function, we can't apply KF as it is meaningless to figure the non-Gaussian function-based mean and the variance. EKF helps in making a non-linear function to be transformed into a linearized function utilizing Taylor Series, it supports having a linear estimate or approximation of a non-linear function.

6.2 IEEE 30 Bus System

The IEEE 30 bus system is considered, same specifications as earlier.

6.3 Performance of Extended Kalman Filter

The Extended Kalman filters, popularly known as EKFs were generally developed for non-linearized discrete time-based processes. It provides an optimal estimate of approximation. The nonlinearities of system-based dynamics are usually estimates in approx. by the linear version of the non-linearized system model around the estimation of the last state. For a valid type of approximation, this process of linearization using the first-order Taylor series should be presented as a good approximation of the non-linearized model in global uncertain domain linked with the estimation of state. The concept of EKFs faces various challenges from the use of linearization process like challenges of implementation, tuning difficulty, issues related to reliability, etc. In the EKFs, the observation models and the transition state, may not be linear functions of a state instead they can be differentiable functions.

- System equations are given as:

The equations of the system are presented as follows:

State equation: $x_{k+1} = f(x_k, u_k) + w_k$ ....1 (a)

Observation equation: $y_k = h(x_k) + v_k$ ....1 (b)

Where,

h = computes the predicted measurement from the predicted state.

f = computes the predicted state from the previous estimate.

However, both the h and f do not apply to the covariance on a direct basis. Rather a partial derivative (the Jacobian) matrix is usually computed. At each step of time, the Jacobian matrix is computed with the present predicted form of states. Such type of matrices is used in the equations of Kalman filter. This type of process necessarily helps to linearize both the current estimate and the non-linearized function. The Kalman filter (KF) represents a method of linear state estimation. On the other hand, the power systems are extremely non-linear nature. For applying the KF to the original state estimation of the power system, an individual needs to perform system linearization. The EKF is generally based on the process of linearizing the non-linear type of equation with the use of Taylor Series where higher-order and quadratic terms are erased. The algorithm in detail is provided by Huang et al (2007) and it presents the following equations:

Prediction: $\hat{x}_{k+1} = f(\hat{x}_k^+) + q_k$ ....(2)

$P_{k+1}^- = F_k P_k^+ F_k^T + Q_k$ ....(3)

Correction: $K_{k+1} = P_{k+1}^- H_{k+1}^T (H_{k+1} P_{k+1}^- H_{k+1}^T + R_{k+1})^{-1}$ ......(4)

$\hat{x}_{k+1} = \hat{x}_{k+1}^- K_{k+1}(z_{k+1} - h(\hat{x}_{k+1}^-))$ ....(5)

$P_{k+1} = (1 - K_{k+1} H_{k+1}) P_{k+1}^-$ ....(6)

$$F_k = \frac{\delta f(x_k)}{\delta x_k}, H_{k+1} = \frac{\delta h(x_{k+1})}{\delta x_{k+1}}$$

EXTENDED KALMAN FILTER

The EKF represents the main methodology that is used to estimate the state of the power system on a current basis. But there are few drawbacks: The Linearization of continuous-discrete (CD) will cause filtering instability if the time interval of calculation is not limited enough.

- For complex systems and large-scale systems, the calculation of the Jacobian matrix is considered to be complicated that is to be executed in case of real-time.
- The disregard of higher and quadratic order terms makes estimation and the prediction less correct/precise.

Figures 6.1 depicts the noise value in a small time framework for EKF.

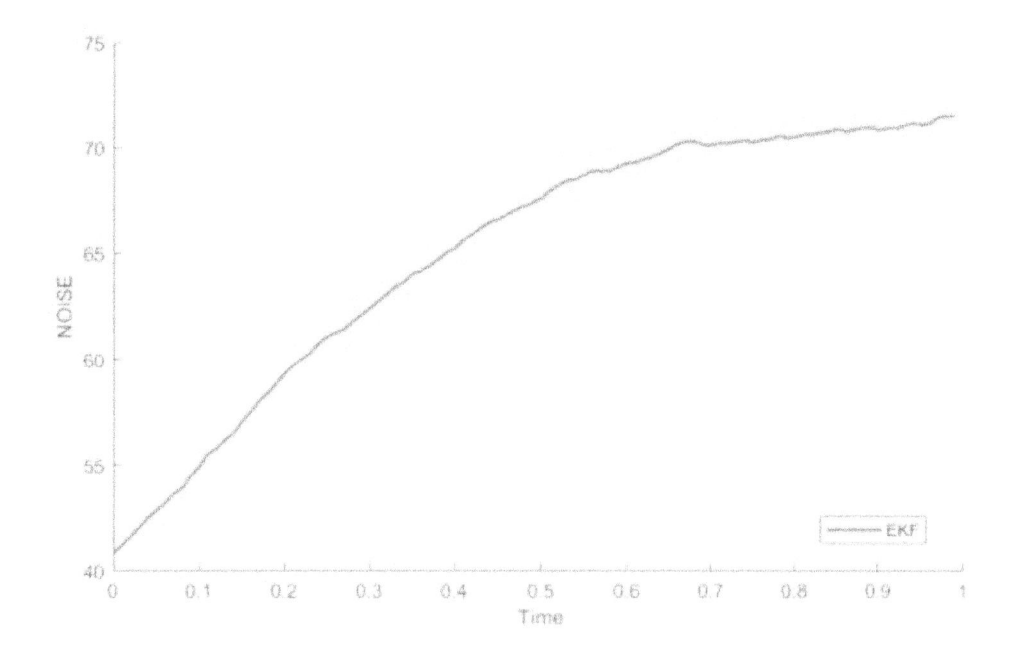

Figure 6.1 EKF noise in a small-time frame

As seen in the graph, the noise increases uniformly, starting at 44 units and maximum noise goes to 72 units in the small time frame of one day.

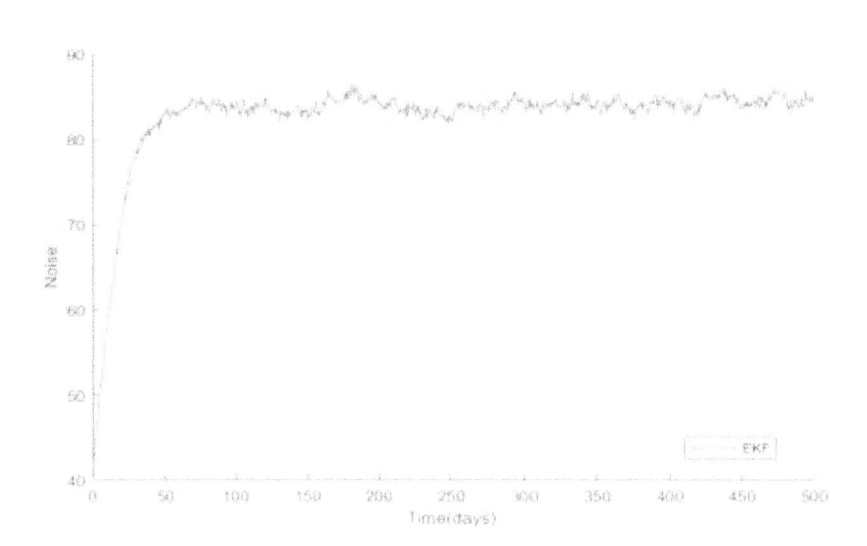

Figures 6.2 depicts the noise value in a large time framework for EKF.

Figure 6.2 EKF noise in a large-time frame

Figures 6.1 and 6.2 depicts the noise value in a small as well as large time framework. Large time involves a 1 day to 500-day analysis process and a small-time frame analyzes different time frames in one day. In a large time frame, noise increases rapidly in the beginning and maximum noise goes to 85 units. In a small time frame maximum noise goes to 70-72 units. EKF only works on non-linear parameter estimation for noise-based reduction.

6.4 Performance of Extended Kalman Filter with Bayesian approach

Figures 6.3 depicts the noise value in a small time framework.

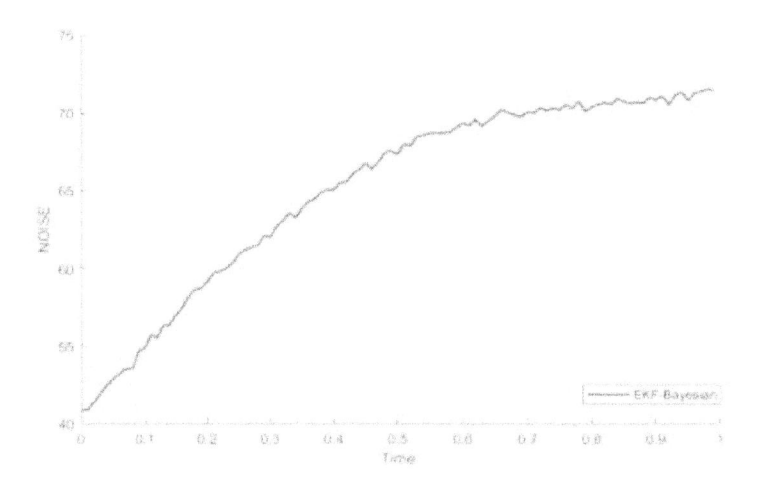

Figure 6.3 EKF-Bayesian noise in a small-time frame

As seen in Fig 6.3, in a small time frame maximum noise goes to 70 units in one day which is less than the EKF in small time frame, hence EKF-Bayesian shows a slight improvement over EKF.

Figures 6.4 depicts the noise value in a large time framework.

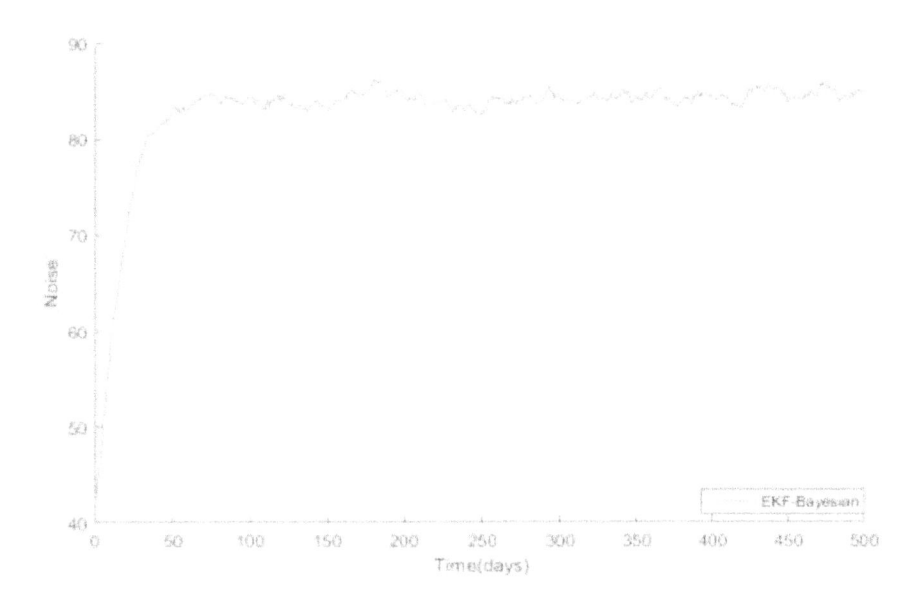

Figure 6.4 EKF-Bayesian noise in a large-time frame

Figures 6.3 and 6.4 depicts the noise value in a small as well as large time framework. In a large time frame, maximum noise goes to 85 units/days. In a small time frame maximum noise goes to 65-70 units. In EKF-Bayesian, non-linearity is improved by Bayesian prediction. Bayesian prediction predicts the non-linearity by noise relation concerning time and PMU's.

6.5 Performance of Extended Kalman Filter with PSO

- It is required that an Extended Kalman Filter to be designed which is based on the Taylor series expansion calculated around a nominal value, subsequently taken as a previous estimate. State transition matrix **F** is then given by the Jacobian vector function $f(\vec{x}, \vec{w})$ about state $\vec{x}$ and the noise scaling matrix τ is given by the Jacobian vector function $f(\vec{x}, \vec{w})$ about state w. As the process dynamics are normally continuous where as the measurements are discrete, hence a hybrid continuous discrete EKF model is to be developed. Here we cannot use the EKF equations of discrete-time and therefore continuous time

EXTENDED KALMAN FILTER

EKF equations have to be arrived at. Also, the measurements being unique, a hybrid using the two is developed and discussed below. Below is explained an observable, non-linear dynamical system, which depicts continuous process dynamics with discrete measurement :

- Here $\vec{x} \in \Re^n$ shows the n-dimensional state vector of the system, $f(.) D_x \to \Re^n$ is a finite non-linear mapping of system states to system inputs, $\vec{z}_k \in D_z \subset \Re^p$ denotes the p-dimensional system measurement, $h(.):D_x \subset \Re^n \to \Re^p$ is a non-linear mapping of system states to output, $\tau_c \in \Re^{n \times w}$ denotes the continuous process noise scaling matrix, $\vec{w} \in D_W \subset \Re^W$ denotes the w-dimensional random process noise and $\vec{v} \in D_v \subset \Re^v$ denotes the v-dimensional random measurement noise.

- System Model of the Extended Kalman Filter optimized Using Particle Swarm Optimization

    - *Predicted state:*
        - $$X_K | X_{K-1} = f(X_{K|K-1, \mu_k}) \quad \quad (8)$$
    - X_K New Prediction
    - X_{K-1} Previous Prediction
    - μ_K variance between prediction
    - *Predicted Covariance:*
        - $$A_{K|K-1} = G_k A_{k|K-1} G_K^T + Q_k \quad \quad (9)$$
    - G_K is covariance matrix
    - Q_K is Linearity in variance
    - **Optimize extended Kalman by PSO**
    - Optimal Kalman
        - $$X_k = A_K | A_{K-1} G_K^T U_K \quad \quad (10)$$
        - $P_i \leftarrow X_K \{X_K \text{ is initial Position of Particle}\} \quad (11)$
    - Initialize the velocity
    - $g \leftarrow P_i \quad \quad (12)$
    - g is random function
    - Update particle Velocity
    - $V_{i,d} = \omega V_{i,d} + \emptyset_p r_p (P_{i,d} - X_{i,d}) + \emptyset_g r_g$
        - $(g_d - X_{i,d}) \quad \quad (13)$
        - $X_i | X_{i-1} = f(x_{i|i-1}, \mu_k) \quad (14)$ From equation (1)

131

EXTENDED KALMAN FILTER

Figures 6.5 depicts the noise levels in a small time framework for EKF-PSO

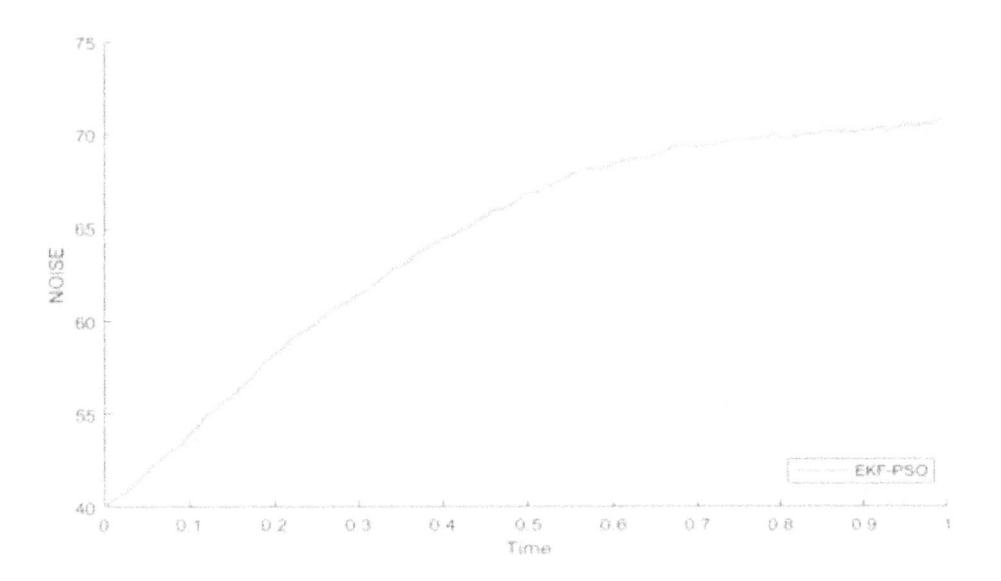

Figure 6.5 EKF-PSO noise in a small-time frame

In fig 6.5, the noise levels in EKF-PSO go upto 70 units in a day which is less as compared to EKF and EKF-Bayesian, hence PSO is able to optimize EKF performance.

Figures 6.6 depicts the noise value in a large time framework for EKF-PSO

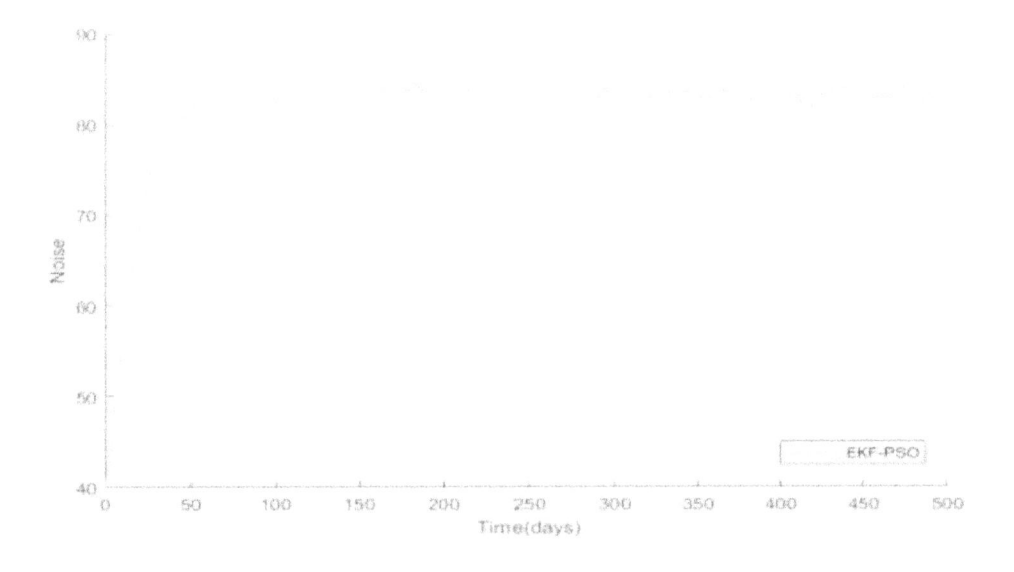

Figure 6.6 EKF-PSO noise in a large-time frame

Large time involves a 1 day to 500-day analysis process and a small-time frame analyzes different time frames in one day. In a large time frame, maximum noise goes to 82 unit. In a small time frame maximum noise goes to 65-70 units. EKF-PSO helps to improve the non-linearity by optimizing local and global parameters. PSO optimizes the non-linear parameters and reduces noise.

6.6 Comparative Performance

6.6.1 Comparative Performance of EKF, EKF-Bayesian and EKF PSO in small time frame

For comparative performance of EKF, EKF-Bayesian and EKF PSO ,we look at fig.6.7 and fig 6.8

Fig 6.7 displays the comparative for small time frame which is summarized in Table 6.1

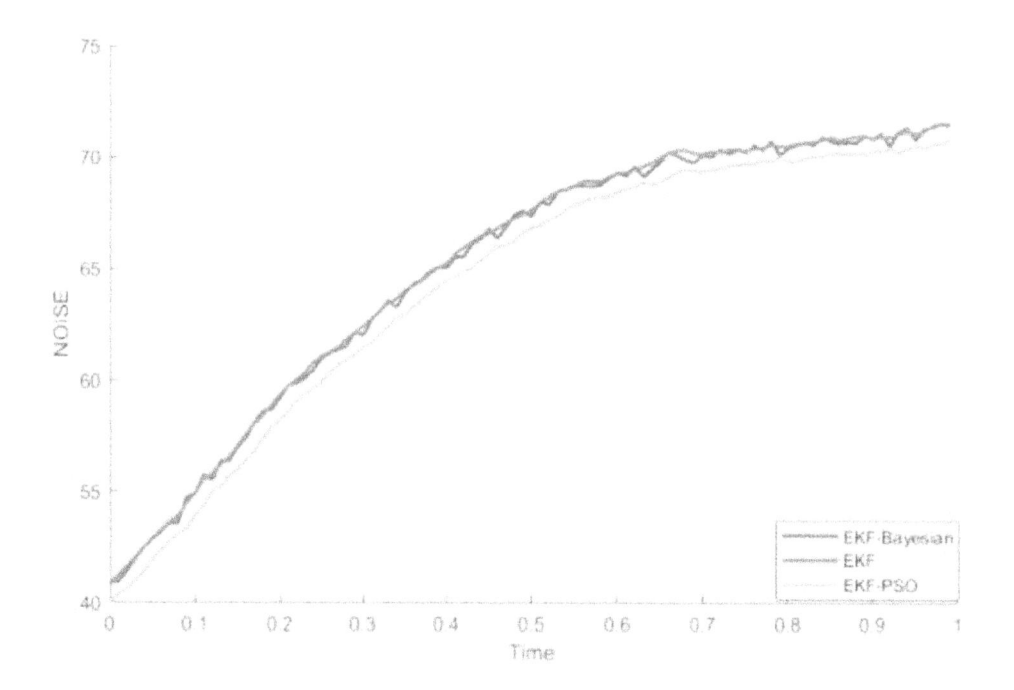

Figure 6.7 Comparison of EKF, EKF-Bayesian and EKF-PSO noise in a small-time frame

It is seen that for EKF PSO, the noise levels are certainly lower than EKF and EKF-Bayesian.

Table 6.1 Comparative of EKF, EKF-Bayesian and EKF-PSO noise in a small-time frame

	Time (Day)	EKF	EKF-Bayesian	EKF PSO
Noise (/kunit)	0.1	53	53	49
	0.2	59	59	58
	0.3	62	62	61
	0.4	65	65	64
	0.5	67	67	66
	0.6	69	69	68
	0.7	70	70	69
	0.8	71	71	70
	0.9	71	71	70

It is seen that for EKF PSO, the noise levels are certainly lower than EKF and EKF-Bayesian at all times of the day. Thus, from this we infer that EKF-PSO helps to improve the non-linearity by optimizing local and global parameters.

6.6.2 Comparative Performance of EKF, EKF-Bayesian and EKF PSO in large time frame

Fig 6.8 displays the comparative of EKF, EKF-Bayesian and EKF PSO for large time frame

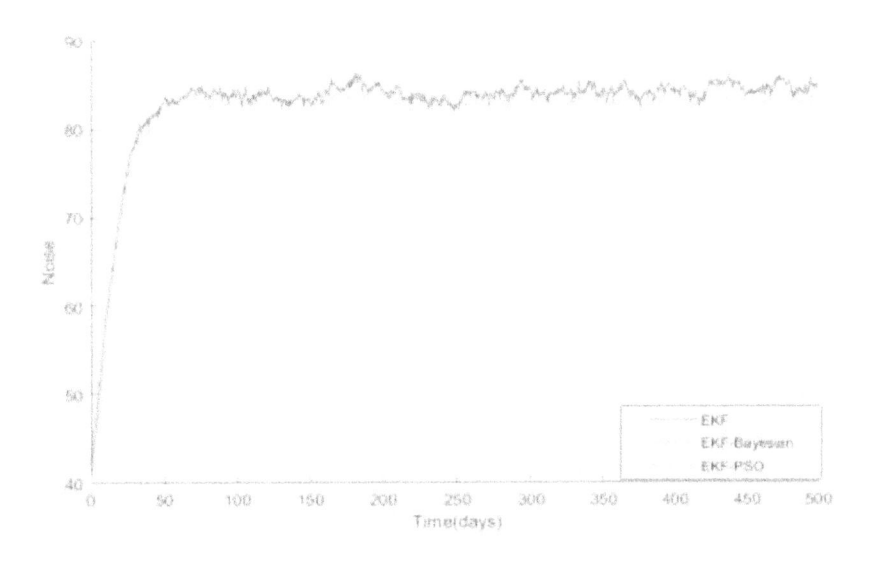

Figure 6.8 Comparison of EKF, EKF-Bayesian and EKF-PSO noise in a large-time frame

As seen in fig 6.8, EKF-PSO depicts minimum noise levels as compared to EKF and EKF-Batesian as it helps to improve the non-linearity by optimization of local and global parameters. The Comparative of the three is also summarized in Table 6.2.

Table 6.2 Comparative Performance of EKF, EKF-Bayesian and EKF-PSO noise in a large time frame

	Time (Days)	EKF	EKF-Bayesian	EKF PSO
Noise (/kunit)	50	82	82	80
	100	84	84	82
	150	84	84	82
	200	86	86	84
	250	84	84	82
	300	84	84	82
	350	84	84	82
	400	84	84	82
	450	84	84	82

As seen the noise after 50 days shows constant values, EKF and EKF-Bayesian remain constant at around 84 units and EKF PSO remains constant at 82 units for different periods.

PSO optimizes the non-linear parameters and reduces noise. EKF-PSO improves noise in small time unit and large time units due to the following reasons.

- First of all, the extended Kalman filter efficiently works on non-linear parameter estimation using the Jacobian matrix.
- In EKF, the parameter is not effectively predicted and optimized. So, it is not able to effectively improve noise when compared to EKF-PSO.
- On comparing EKF-Bayesian to EKF-PSO, EKF-PSO improves noise because of optimizing the predicted parameters or in other words refine the parameters iteratively.

6.6.3 Comparative Performance of EKF, Bayesian -Kalman and PSO-Kalman

In fig 6.9, we have compared the performance of Extended Kalman filter, Bayesian-Kalman filter and PSO -Kalman filter performances.

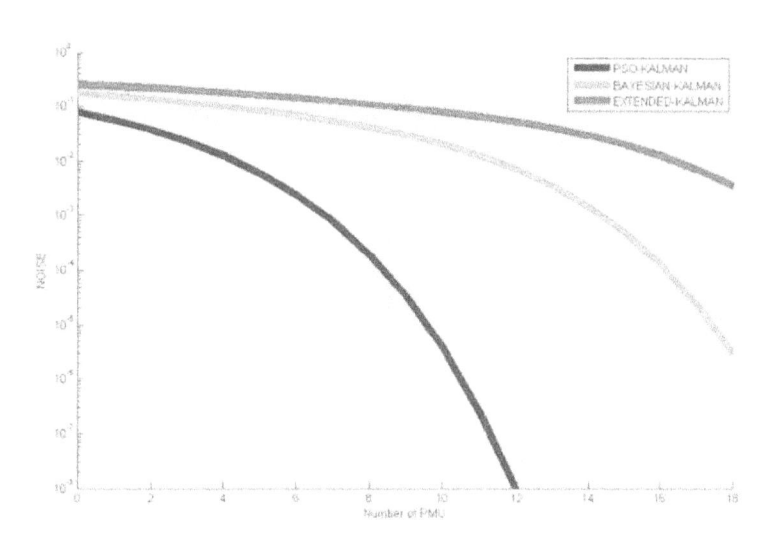

Figure 6.9 Comparison of Noise in PSO-Kalman, Bayesian-Kalman and EKF

Figure 6.9 depicts the noise level according to increasing resources(PMU). Here the Red line is for the Extended Kalman filter, the Green line is for the Bayesian -Kalman and the Blue line is for PSO-Kalman. The figure represents the noise level of the EKF, PSO-Kalman and Bayesian-Kalman. Here we see that the curves in the EKF and Bayesian-Kalman show a high level of noise as compared to PSO-Kalman.

EXTENDED KALMAN FILTER

6.6.4 Comparative Performance of EKF, EKF-Bayesian and KF-Bayesian

In fig 6.10, we have compared the performance of Extended Kalman filter, Bayesian-Kalman filter and Bayesian-Extended Kalman filter performances.

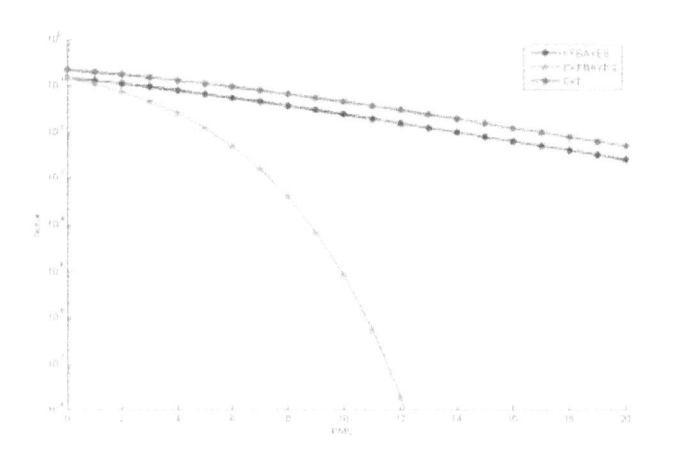

Figure 6.10 Comparison of Noise in KFBAYES, EKFBAYES, EKF

Figure 6.10 depicts a comparison of three different approaches for different number of PMUs. Here, the purple line represents the Extended Kalman filter, blue line representing the Kalman filter with the Bayesian form and sky blue Line represents the Extended Kalman filter with the Bayesian. In the graph, the curve for the Extended Kalman filter depicts decreased noise and stability where as the curve of the Kalman filter with Bayesian depicts a high noise level. In this part, different simulation results have been evaluated based on the proposed EKF on the IEEE-30 bus system using MATLAB software. The analysis is based on noise.

6.6.5 Comparative Performance for different Kalman filters and various approaches

In this part, the overall average performance of different approaches showing comparison for two parameters, noise and MSE. In figure 6.11 and table 6.3 depict the analysis based on noise for Kalman, EKF, UKF, Kalman-Taylor with PSO, Kalman Bayesian with PSO, Kalman-Bayesian-Taylor.

Table 6.3 Analysis of noise comparison in 30 bus system

Bus system	Kalman	EKF	UKF	Kalman-Taylor	Kalman-bayesian	Kalman-bayesian-Taylor
30-bus	0.6	0.62	0.76	0.55	0.42	0.2

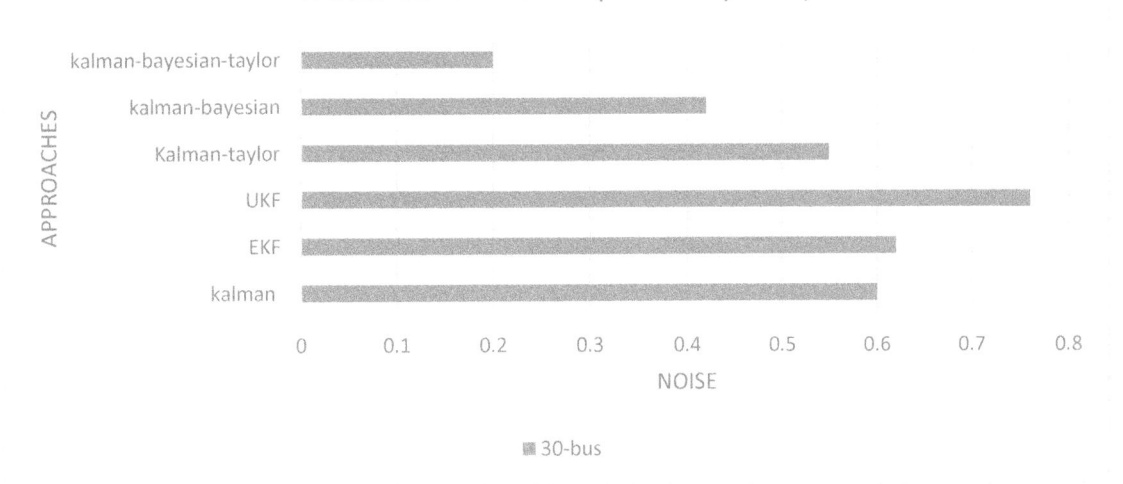

Figure 6.11 Analysis of noise comparison in 30-bus system

As seen earlier, Kalman filter gives better performance compared to its extensions, namely UKF and EKF but then Kalman is not for non-linear systems for which UKF and EKF are better but tedious calculations. Therefore, Kalman filter is linearized and prediction improved using combination of Kalman with Taylor expansion and Bayesian prediction. It is now observed that Kalman-Taylor-Bayesian improves noise as compared to other approaches because prediction and optimization work parallelly.

The overall average performance of different approaches is compared concerning the Mean Square Error also. Figure 6.12 and Table 6.4 analyze the MSE for Kalman, EKF, UKF, Kalman-Taylor with PSO, Kalman Bayesian with PSO, Kalman-Bayesian-Taylor.

Table 6.4 Analysis of MSE comparison in 30-bus system

Bus system	Kalman	EKF	UKF	Kalman-Taylor	Kalman-bayesian	Kalman-bayesian-Taylor
30-bus	37.2	47.2	41.8	23.1	10.12	8.6

Here again, it is seen that Kalman-Taylor-Bayesian improves MSE as compared to other approaches.

EXTENDED KALMAN FILTER

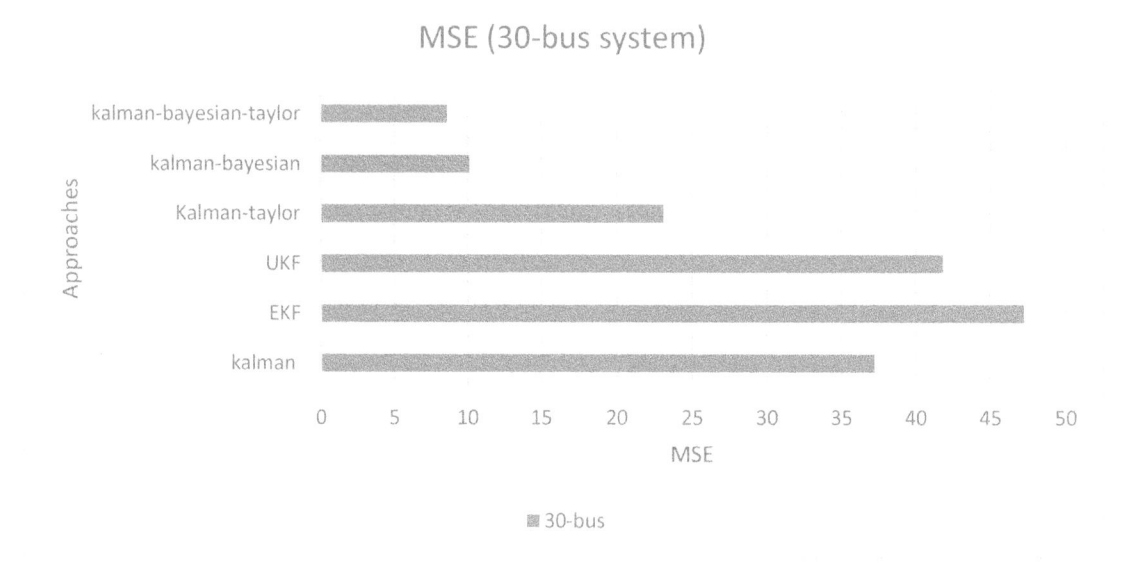

Figure 6.12 Analysis of noise comparison in 30 -bus system

The MSE is minimum, which is around 7.5 and maximum is for EKF around 47, followed by UKF, Kalman, Kalman-Taylor, Kalman-Bayesian. It is again observed that Kalman-Taylor-Bayesian improves noise as compared to other approaches because prediction and optimization work parallelly.

6.7 Conclusion

As Kalman filter was applicable to linear systems and noise being nonlinear, the extended Kalman filter which works on nonlinear systems was explored. Extended Kalman filter was combined with the Bayesian prediction and a hybrid with PSO was also studied. Comparison of various approaches was done. It finely concluded that the EKF-PSO improves the noise in a small time and large time frame.

Here, a comparison of EKF was done with some previously explored approaches too. In fig 6.9, we have compared the performance of Extended Kalman filter, Bayesian-Kalman filter and PSO -Kalman filter performances and as per the result Kalman-PSO performs best. In fig 6.10, we have compared the performance of Extended Kalman filter, Bayesian-Kalman filter and Bayesian-Extended Kalman filter performances, in this case Bayesian-Extended Kalman filter performs best. Performance is based on the ability to reduce noise.

In this chapter, comparative analysis of different parameters like noise and MSE is done in for the 30 bus system . Figure 6.11 analyzes the noise in different numbers of PMU based on different approaches like Kalman filter, UKF, EKF, Kalman-Taylor, Kalman-Bayesian, Kalman-Bayesian -Taylor approaches. Fig. 6.12 depicts MSE for different filters and approaches. The results show that the hybrid approach improves MSE and noise as compared to standalone approaches.

REFERENCES

[1] Ageev, A., Bortalevich, S., Loginov, E., Shkuta, A., & Sorokin, D. (2018). There is need in new generation smart grid for the space and ground energy systems. In *MATEC Web of Conferences* (Vol. 158, p. 01001). EDP Sciences.

[2] Sharma, A., & Samantaray, S. R. (2018). Power system tracking state estimator for smart grid under unreliable PMU data communication network. *IEEE Sensors Journal*, *18*(5), 2107-2116.

[3] Tayal, Nisha & Rintu Khanna. (2018). A novel approach of parameter estimation of smart grid by Kalman Filter with Taylor Expansion. *Global Journal of Engineering Science and Reseachers.* 5(9), 2348-8034.

[4] Kong, X., Chen, Y., Xu, T., Wang, C., Yong, C., Li, P., & Yu, L. (2018). A hybrid state estimator based on SCADA and PMU measurements for medium voltage distribution system. *Applied Sciences*, *8*(9), 1527.

[5] Yang, H., Zhang, W., Chen, J., & Wang, L. (2018). PMU-based voltage stability prediction using least square support vector machine with online learning. *Electric Power Systems Research*, *160*, 234-242.

[6] Kurt, M. N., Yılmaz, Y., & Wang, X. (2018). Distributed quickest detection of cyber-attacks in smart grid. *IEEE Transactions on Information Forensics and Security*, *13*(8), 2015-2030.

[7] Theodorakatos, N. P. (2018). Optimal phasor measurement unit placement for numerical observability using a two-phase branch-and-bound algorithm. *International Journal of Emerging Electric Power Systems*, *19*(3).

[8] Shi, Y., Tuan, H. D., Nasir, A. A., Duong, T. Q., & Poor, H. V. (2018). PMU Placement Optimization for Smart Grid Observability and State Estimation. *arXiv preprint arXiv:1806.02541*.

[9] Cui, S., Yu, Q., Gu, G., & Gang, Q. (2017, November). Research on the architecture of electric power information communication network for smart grid. In *Energy Internet and Energy System Integration (EI2), 2017 IEEE Conference on* (pp. 1-4). IEEE.

[10] Molina, M. G. (2017). Energy storage and power electronics technologies: a strong combination to empower the transformation to the smart grid. *Proceedings of the IEEE*, *105*(11), 2191-2219.

REFERENCES

[11] Lascu, C. E. (2017, October). HOW TO IMPLEMENT A SMART GRID? In *International Conference on Management and Industrial Engineering* (No. 8, pp. 36-43). Niculescu Publishing House.

[12] Yoldaş, Y., Önen, A., Muyeen, S. M., Vasilakos, A. V., & Alan, İ. (2017). Enhancing smart grid with microgrids: Challenges and opportunities. *Renewable and Sustainable Energy Reviews*, *72*, 205-214.

[13] Hou, L., & Wang, C. (2017, October). Market-based mechanisms for smart grid management: Necessity, applications and opportunities. In *Systems, Man, and Cybernetics (SMC), 2017 IEEE International Conference on* (pp. 2613-2618). IEEE.

[14] Cintuglu, M. H., Mohammed, O. A., Akkaya, K., & Uluagac, A. S. (2017). A Survey on Smart Grid Cyber-Physical System Testbeds. *IEEE Communications Surveys and Tutorials*, *19*(1), 446-464.

[15] Akhlaghi, S., Zhou, N., & Huang, Z. (2017, July). Adaptive adjustment of noise covariance in Kalman filter for dynamic state estimation. In *2017 IEEE Power & Energy Society General Meeting* (pp. 1-5). IEEE.

[16] Maji, T. K., & Acharjee, P. (2017). Multiple solutions of optimal PMU placement using exponential binary PSO algorithm for smart grid applications. *IEEE Transactions on Industry Applications*, *53*(3), 2550-2559.

[17] Rahman, N. H. A., & Zobaa, A. F. (2017). Integrated mutation strategy with modified binary PSO algorithm for optimal PMUs placement. *IEEE Transactions on Industrial Informatics*, *13*(6), 3124-3133.

[18] McCamish, B., Kulkarni, J., Ke, Z., Harpool, S., Huo, C., Brekken, T., ... & Yokochi, A. (2017). A rapid PMU-based load composition and PMU estimation method. *Electric Power Systems Research*, *143*, 44-52.

[19] Yang, D., Nie, Z., Jones, K., & Centeno, V. (2017, September). Adaptive decision-trees-based regional voltage control. In *2017 North American Power Symposium (NAPS)* (pp. 1-6). IEEE.

[20] Chui, Charles K., and Guanrong Chen. *Kalman filtering*. Springer International Publishing, 2017.

[21] Galád, M., Špánik, P., Cacciato, M., & Nobile, G. (2017). Analysis of state of charge estimation methods for smart grid with VRLA batteries. *Electrical Engineering*, *99*(4), 1233-1244.

REFERENCES

[22] Zhao, J., & Mili, L. (2017). A framework for robust hybrid state estimation with unknown measurement noise statistics. *IEEE Transactions on Industrial Informatics, 14*(5), 1866-1875.

[23] Noureen, S. S., Roy, V., & Bayne, S. B. (2017, December). Phasor measurement unit integration: A review on optimal PMU placement methods in power system. In *2017 IEEE Region 10 Humanitarian Technology Conference (R10-HTC)* (pp. 328-332). IEEE.

[24] Musleh, A. S., Muyeen, S. M., Al-Durra, A., & Khalid, H. M. (2016, November). PMU based wide area voltage control of smart grid: A real time implementation approach. In *Innovative Smart Grid Technologies-Asia (ISGT-Asia), 2016 IEEE* (pp. 365-370). IEEE.

[25] Shahriar, A. Z. M., Taylor, G. A., & Bradley, M. E. (2016, September). Parameter estimation and sensitivity analysis of distribution network equivalents. In *Power Engineering Conference (UPEC), 2016 51st International Universities* (pp. 1-6). IEEE.

[26] Rauf, S., Rasool, S., Rizwan, M., Yousaf, M., & Khan, N. (2016). Domestic electrical load management using smart grid. *Energy Procedia, 100*, 253-260.

[27] Nur, A., & Kaygusuz, A. (2016, April). Load control techniques in smart grids. In *Smart Grid Congress and Fair (ICSG), 2016 4th International Istanbul* (pp. 1-4). IEEE.

[28] Wang, X., & Yaz, E. E. (2016). Smart power grid synchronization with fault tolerant nonlinear estimation. *IEEE Transactions on Power Systems, 31*(6), 4806-4816.

[29] Rawat, A., & Joshi, S. N. (2015). -Phasor Measurement Unit in Smart Grid for Minimum Elapsed Time. *International Journal on Recent and Innovation Trends in Computing and Communication, 3*(10), 5842-5844.

[30] Gore, R., & Kande, M. (2015, March). Analysis of wide area monitoring system architectures. In *Industrial Technology (ICIT), 2015 IEEE International Conference on* (pp. 1269-1274) IEEE.

[31] Larik, R. M., Mustafa, M. W., & Qazi, S. H. (2015). Smart grid technologies in power systems: an overview. *Research Journal of Applied Sciences, Engineering and Technology, 11*(6), 633-638.

[32] Schellong, W., & Gerngross, S. (2015, November). Energy demand analysis in smart grids. In *Energy and Sustainability Conference (IESC), 2015 International* (pp. 1-6). IEEE.

REFERENCES

[33] Torrent-Fontbona, F. (2015). Optimization methods meet the smart grid. New methods for solving location and allocation problems under the smart grid paradigm.

[34] Baldwin, T. L., L. Mili, M. B. Boisen, and R. Adapa,"Power system observability with minimal phasor measurement placement." *IEEE Transactions on Power Systems*, Vol 8, No. 2, pp: 707-715, 1993

[35] Amin, Massoud. (2015) "Smart Grid." *PUBLIC UTILITIES FORTNIGHTLY*.

[36] Zhenyu Huang, Kevin Schneider, Ning Zhou and Jarek Nieplocha, "Estimating power system dynamic states using extended Kalman filter", IEEE PES General meeting: conference and exposition, pp.1–5, 2014. DOI: 10.1109/PESGM.2014.6939934

[37] Gupta, D. K., & Pandey, R. K. (2014, December). Grid stabilization with PMU signals—A survey. In *Power Systems Conference (NPSC), 2014 Eighteenth National* (pp. 1-6). IEEE.

[38] Sirisukprasert, S. (2014, March). Power electronics-based energy storages: A key component for Smart Grid technology. In *Electrical Engineering Congress (iEECON), 2014 International* (pp. 1-7). IEEE.

[39] Gupta, D. K., & Pandey, R. K. (2014, December). Grid stabilization with PMU signals— A survey. In *Power Systems Conference (NPSC), 2014 Eighteenth National* (pp. 1-6). IEEE.

[40] Reddy, K. S., Kumar, M., Mallick, T. K., Sharon, H., & Lokeswaran, S. (2014). A review of Integration, Control, Communication and Metering (ICCM) of renewable energy based smart grid. *Renewable and Sustainable Energy Reviews*, 38, 180-192.

[41] Colak, I., Bayindir, R., Fulli, G., Tekin, I., Demirtas, K., & Covrig, C. F. (2014). Smart grid opportunities and applications in Turkey. *Renewable and Sustainable Energy Reviews*, 33, 344-352.

[42] Lee, S. H. (2014). Real-time camera tracking using a particle filter combined with unscented Kalman filters. *Journal of Electronic Imaging*, 23(1), 013029.

[43] Moldes, E. R. (2013). *Flexible load management in Smart grids* (Doctoral dissertation, Master Thesis, Aalborg Universitet).

[44] Ahat, M., Amor, S. B., Bui, M., Bui, A., Guérard, G., & Petermann, C. (2013). Smart grid and optimization. *American Journal of Operations Research*, 3(01), 196.

[45] Yan, H., Huang, G., Wang, H., & Shu, R. (2013, December). Application of unscented kalman filter for flying target tracking. In *Information Science and Cloud Computing (ISCC), 2013 International Conference on* (pp. 61-66). IEEE.

REFERENCES

[46] Sutar, C., Verma, D. K., Amethi, R., & Sultapur, K. (2013). Application of phasor measurement unit in smart grid. *Pratibha: International Journal of Science, Spirituality, Business and Technology (IJSSBT)*, *1*(2).

[47] Acharjee, P. (2013). Strategy and implementation of Smart Grids in India. *Energy Strategy Reviews*, *1*(3), 193-204.

[48] Laverty, D. M., Best, R. J., Brogan, P., Al Khatib, I., Vanfretti, L., & Morrow, D. J. (2013). The Open PMU platform for open-source phasor measurements. *IEEE Transactions on Instrumentation and Measurement*, *62*(4), 701-709.

[49] Ma, R., Chen, H. H., Huang, Y. R., & Meng, W. (2013). Smart grid communication: Its challenges and opportunities. *IEEE transactions on Smart Grid*, *4*(1), 36-46.

[50] Huang, Y. F., Werner, S., Huang, J., Kashyap, N., & Gupta, V. (2012). State estimation in electric power grids: Meeting new challenges presented by the requirements of the future grid. *IEEE Signal Processing Magazine*, *29*(5), 33-43.

[51] Paramanand, C., & Rajagopalan, A. N. (2012). Depth from motion and optical blur with an unscented Kalman filter. *IEEE Transactions on Image Processing*, *21*(5), 2798-2811.

[52] Zhao, Y., Chen, W., & Gao, J. (2012). An improved UKF algorithm and its performance on target tracking based on electronic technology. In *Advances in Mechanical and Electronic Engineering* (pp. 387-391). Springer, Berlin, Heidelberg.

[53] Ding, Q., Zhao, X., & Han, J. (2012, June). Adaptive unscented Kalman filters applied to visual tracking. In *Information and Automation (ICIA), 2012 International Conference on* (pp. 491-496). IEEE.

[54] Kavaiya, Madhavi, Kartik Pandya, "PMU Placement for Power System Observability using Integer Linear Programming" *International Journal of Engineering Development And Research, IJEDR, Vadodara, India*, Vol. 22, No. 1, 2007.

[55] Siddharth Deshmukh, Balasubramaniam Natarajan and Anil Pahwa, "Stochastic State Estimation for Smart Grids in the Presence of Intermittent Measurements", IEEE Latin-America Conference on Communications (LATINCOM), pp. 1–6, 2012. DOI: 10.1109/LATINCOM.2012.6506013.

[56] Wang, Q., & Xiao, D. (2012, August). GPS/SINS positioning method based on robust UKF. In *Industrial Control and Electronics Engineering (ICICEE), 2012 International Conference on* (pp. 877-881). IEEE.

REFERENCES

[57] Rihan, M., Ahmad, M., & Beg, M. S. (2011, December). Phasor measurement units in the Indian smart grid. In *Innovative Smart Grid Technologies-India (ISGT India), 2011 IEEE PES* (pp. 261-267). IEEE.

[58] E. Ghaheramani and I. Kamwa, "Dynamic state estimation in power system by applying the extended Kalman filter with unknown inputs to phasor measurements", IEEE Transactions Power Systems, Vol.26, No. 4, pp. 2556–2566, 2011. DOI: 10.1109/TPWRS.2011.2145396.

[59] Gomez-Exposito, A., Abur, A., de la Villa Jaen, A., & Gomez-Quiles, C. (2011). A multilevel state estimation paradigm for smart grids. *Proceedings of the IEEE, 99*(6), 952-976.

[60] Galli, S., Scaglione, A., & Wang, Z. (2011). For the grid and through the grid: The role of power line communications in the smart grid. *Proceedings of the IEEE, 99*(6), 998-1027.

[61] Rihan, M., Ahmad, M., & Beg, M. S. (2011, December). Phasor measurement units in the Indian smart grid. In *Innovative Smart Grid Technologies-India (ISGT India), 2011 IEEE PES* (pp. 261-267). IEEE.

[62] Patel, A., Aparicio, J., Tas, N., Loiacono, M., & Rosca, J. (2011, October). Assessing communications technology options for smart grid applications. In *Smart Grid Communications (SmartGridComm), 2011 IEEE International Conference on* (pp. 126-131). IEEE.

[63] Amin, M. M., Moussa, H. B., & Mohammed, O. A. (2011, August). Development of a wide area measurement system for smart grid applications. In *This paper has been presented in the 18th World Congress of the International Federation of Automatic Control (IFAC) Invited Session, Milano, Italy, Aug.*

[64] Hayashiya, H., Yoshizumi, H., Suzuki, T., Furukawa, T., Kondoh, T., Kitano, M., ... & Miyagawa, T. (2011, August). Necessity and possibility of smart grid technology application on railway power supply system. In *Power Electronics and Applications (EPE 2011), Proceedings of the 2011-14th European Conference on* (pp. 1-10). IEEE.

[65] Yu, R., Zhang, Y., Gjessing, S., Yuen, C., Xie, S., & Guizani, M. (2011). Cognitive radio based hierarchical communications infrastructure for smart grid. *IEEE network, 25*(5).

[66] Farhangi, H. (2010). The path of the smart grid. *IEEE power and energy magazine, 8*(1).

REFERENCES

[67] Overbye, T. J., & Weber, J. D. (2010, July). The smart grid and PMUs: Operational challenges and opportunities. In *Power and Energy Society General Meeting, 2010 IEEE* (pp. 1-5). IEEE.

[68] Moslehi, K., & Kumar, R. (2010). A reliability perspective of the smart grid. *IEEE Transactions on Smart Grid, 1*(1), 57-64.

[69] SmartGrid:https://www.nema.org/Policy/Energy/Smartgrid/Pages/default.aspx accessed on 5/12/2018 at 1.00 PM.

[70] Smart Grid Evolution: http://www.eolasmagazine.ie/smart-grid-evolution/ accessed on 5/12/2018 at 1.15 PM.

[71] Smart Grid Technology: https://www.esi-africa.com/power-of-smart-grid-technology/accessed on 6/12/2018 at 3.00 PM.

[72] Mamata Madhumita and Soumya Ranjan Aich, 2010, 'An Approach to Design a Power System Harmonic Estimator', Bachelor in Technology in Electrical Engineering, National Institute of Technology, Rourkela.

[73] Considine, T., Principal, T. C., & Cox, W. T. (2009). Smart Loads and Smart Grids—Creating the Smart Grid Business Case. *Presented at GridInterop*, 1.

[74] Diao, R., Sun, K., Vittal, V., O'Keefe, R. J., Richardson, M. R., Bhatt, N., ... & Sarawgi, S. K. (2009). Decision tree-based online voltage security assessment using PMU measurements. *IEEE Transactions on Power Systems, 24*(2), 832-839.

[75] Peng, Chunhua, and Xuesong Xu,"A hybrid algorithm based on BPSO and immune mechanism for PMU optimization placement." *Intelligent Control and Automation, 2008. WCICA 2008. 7th World* Congress on, pp. 7036-7040. IEEE, 2008..

[76] Gou, B. (2008). Generalized integer linear programming formulation for optimal PMU placement. *IEEE transactions on Power Systems, 23*(3), 1099-1104.

[77] Huang, Z., Schneider, K., & Nieplocha, J. (2007, December). Feasibility studies of applying Kalman filter techniques to power system dynamic state estimation. In *Power Engineering Conference, 2007. IPEC 2007. International* (pp. 376-382). IEEE.

[78] Ashwani Kumar, Biswaroop Das and Jaydev Sharma, "Robust Dynamic State Estimation of Power System Harmonics", International Journal of Electrical Power and Energy Systems, Vol.28, pp. 65–74, 2006.

[79] Zhao, L., & Abur, A. (2005). Multi area state estimation using synchronized phasor measurements. *IEEE Transactions on Power Systems, 20*(2), 611-617.

[80] Mohammadi-Ivatloo, B., and S. H. Hosseini, "Optimal PMU placement for power system observability considering secondary voltage control." *In Electrical and*

REFERENCES

Computer Engineering, 2008. CCECE 2008. Canadian Conference on, pp. 000365-000368. IEEE, 2008..

[81] Al-Othman, A. K., & Irving, M. R. (2005). Uncertainty modelling in power system state estimation. *IEE Proceedings-Generation, Transmission and Distribution, 152*(2), 233-239.

[82] Moscato, P., Cotta, C., & Mendes, A. (2004). Memetic algorithms. In *New optimization techniques in engineering* (pp. 53-85). Springer, Berlin, Heidelberg.

[83] VanDyke, M. C., Schwartz, J. L., & Hall, C. D. (2004). Unscented Kalman filtering for spacecraft attitude state and parameter estimation. *Advances in the Astronautical Sciences, 118*(1), 217-228.

[84] Julier, S. J., & Uhlmann, J. K. (2004). Unscented filtering and nonlinear estimation. *Proceedings of the IEEE, 92*(3), 401-422.;

[85] Blum, C., & Roli, A. (2003). Metaheuristics in combinatorial optimization: Overview and conceptual comparison. *ACM computing surveys (CSUR), 35*(3), 268-308.

[86] Milosevic, B., & Begovic, M. (2003). Voltage-stability protection and control using a wide-rea network of phasor measurements. *IEEE Transactions on Power Systems, 18*(1), 121-127.

[87] Mitchell, J. E. (2002). Branch-and-cut algorithms for combinatorial optimization problems. *Handbook of applied optimization*, 65-77.

[88] Yu, C. S., Liu, C. W., Yu, S. L., & Jiang, J. A. (2002). A new PMU-based fault location algorithm for series compensated lines. *IEEE Transactions on Power Delivery, 17*(1), 33-46.

[89] SreenivasaReddy, P. S., S. P. Chowdhury, and S. Chowdhury, "PMU placement-a comparative survey and review." *In Developments in Power System Protection (DPSP 2010). Managing the Change, 10th IET International Conference on*, IET, pp. 1-4, 2010.

[90] Yang, J. Z., & Liu, C. W. (2001). A precise calculation of power system frequency. *IEEE Transactions on Power Delivery, 16*(3), 361-366.

[91] Jiang, J. A., Yang, J. Z., Lin, Y. H., Liu, C. W., & Ma, J. C. (2000). An adaptive PMU based fault detection/location technique for transmission lines. I. Theory and algorithms. *IEEE Transactions on Power Delivery, 15*(2), 486-493.

[92] Sinha A K, Mandal J K. Dynamic State Estimation Using ANN based Bus Load Prediction [J]. IEEE Trans on Power Systems, 1999, 14(11):1219-1225.

REFERENCES

[93] MAO Yuhua, et al. Adaptive Kalman Filter Method for State Estimation in Power System Journal of Northeast China Institute of Electric Power Engineering, 1995, 15(2):20-26.

[94] Schnabel, R. B., & Eskow, E. (1990). A new modified Cholesky factorization. *SIAM Journal on Scientific and Statistical Computing*, *11*(6), 1136-1158.

[95] Stentz, A. (1994, May). Optimal and efficient path planning for partially-known environments. In *ICRA* (Vol. 94, pp. 3310-3317).

[96] Chakrabarti, Saikat, Elias Kyriakides, and Demetrios G. Eliades, "Placement of synchronized measurements for power system observability." *IEEE Transactions on Power Delivery*, Vol 24, No. 1 pp: 12-19, 2009.

[97] K. Nishiya, J. Hasegawa and T. Koike, "Dynamic state estimation including anomaly detection and identification for power system", IEE Proceedings Gen. Trans. and Dist., Vol.129, No.5, pp. 192–198, 1982. DOI: 10.1049/ip-c.1982.0032

[98] Hurtgen, Michaël, and J-C. Maun, "Optimal PMU placement using iterated local search." *International journal of electrical power & energy systems*, Brussels, Belgium, Vol. 32, No. 8 pp: 857-860, 2010.

[99] Narendra, P. M., & Fukunaga, K. (1977). A branch and bound algorithm for feature subset selection. *IEEE Transactions on computers*, (9), 917-922.

[100] Yuill, William, A. Edwards, S. Chowdhury, and S. P. Chowdhury, "Optimal PMU placement: A comprehensive literature review." *In Power and Energy Society General Meeting, IEEE*, pp. 1-8. IEEE, 2011.

[101] Sorenson, H. W. (1970). Least-squares estimation: from Gauss to Kalman. *IEEE spectrum*, *7*(7), 63-68.

[102] Schweppe, F. C., & Wildes, J. (1970). Power system static-state estimation, Part I: Exact model. *IEEE Transactions on Power Apparatus and systems*, (1), 120-125.

[103] Hart, P. E., Nilsson, N. J., & Raphael, B. (1968). A formal basis for the heuristic determination of minimum cost paths. *IEEE transactions on Systems Science and Cybernetics*, *4*(2), 100-107.

[104] Smart Grid: Advanced Monitoring: http://www.indiasmartgrid.org/Advanced-Metering-Infrastructure.php accessed on 6/12/2018 at 5.00 PM.

[105] DistributedSmartGrid:https://www.smartgrid.gov/the_smart_grid/distribution_intellige nce.html accessed on 7/12/2018 at 5.00 PM.

[106] Manousakis, N. M., G. N. Korres, and P. S. Georgilakis, "Optimal placement of phasor measurement units: A literature review." *In Intelligent System Application to Power*

REFERENCES

Systems (ISAP), 2011 16th International Conference on, IEEE, Athens, Greece, pp. 1-6. 2011.

[107] Smart Grid Evolution: A New Generation of Intelligent Electronic Devices: https://www.renewableenergyworld.com/articles/2014/03/smart-grid-evolution-a-new-generation-of-intelligent-electronic-devices.html accessed on 7/12/2018 at 6.00 PM.

[108] Smart Grid: https://www.techopedia.com/definition/692/smart-grid accessed on 21/12/2018 at 6.00 PM.

[109] Smart Grid: http://www.powerful-thinking.org.uk/factsheet/smart-grids-explained/ accessed on 21/12/2018 at 6.00 PM.

[110] [110] Smart grid concepts: http://large.stanford.edu/courses/2011/ph240/bogdanowicz1/ accessed on 21/12/2018 at 5.00 PM.

[111] Cruz, Marco ARS, Helder RO Rocha, Marcia HM Paiva, Marcelo EV Segatto, Eglantine Camby, and Gilles Caporossi, "An algorithm for cost optimization of PMU and communication infrastructure in WAMS." *International Journal of Electrical Power & Energy Systems*, Binghamton, United States, Vol. 106, pp: 96-104, 2019.

[112] Sarailoo, Morteza, N. Eva Wu, and John S. Bay, "Toward a spoof-tolerant PMU network architecture." *International Journal of Electrical Power & Energy Systems,* Binghamton, United States Vol.107, pp: 311-320.

[113] Al Rammal, Zeina, Nivine Abou Daher, Hadi Kanaan, Imad Mougharbel, and Maarouf "Optimal PMU placement for reverse power flow detection." In *2018 4th International Conference on Renewable Energies for Developing Countries (REDEC)*, IEEE, Hadath, Lebanon, pp. 1-5, 2018.

[114] Jinghe Zhang, Welch, N.Ramakrishnan and Saifur Rahman, "Kalman Filters for Dynamic and Secure Smart Grid State Estimation." *Intelligent Industrial Systems,* Vol. 1, pp: 29-36, 2015.

[115] Xin Wang and Edwin E. Yaz, "Smart Power Grid Synchronization with Fault Tolerant Nonlinear Estimation." *IEEE Transactions on Power Systems* Volume 31, Issue 6, 2016.

[116] Zhu, Xingzheng, Miles HF Wen, Victor OK Li, and Ka-Cheong Leung, "Optimal PMU-Communication Link Placement for Smart Grid Wide-Area Measurement Systems." *IEEE Transactions on Smart Grid*, pp: 1949-3053, 2018.

REFERENCES

[117] Junbo Zhao, Marcos Netto, Lamine Mili, "A Robust Iterated Extended Kalman Filter for Power System Dynamic State Estimation." *IEEE Power and Energy Society General Meeting (PESGM)* (2016)

[118] Hadis Karimipour, "Extended Kalman Filter Based Parallel Dynamic State Estimation." *IEEE Transactions on Smart Grid*, Vol. 6, Issue 3, 2015

[119] Jinghe Zhang, Greg Welch, Gary Bishop and Zhenyu Hua "A Two Stage Kalman Filter approach for Robust and Real Time Power System State Estimation." *IEEE Transactions on Sustainable Energy*, Vol 5, Issue 2, 2014)

[120] Rahman, Nadia Hanis Abd, and Ahmed Faheem Zobaa, "Integrated Mutation Strategy with Modified Binary PSO Algorithm for Optimal PMUs Placement." *IEEE Transactions on Industrial Informatics*, Vol 13, No.6, pp: 3124-3133, 2017.

[121] Kumari, Saroj, Pratima Walde, Asif Iqbal, and Akash Tyagi, "Optimal phasor measuring unit placement by binary particle swarm optimization." In *Computing, Communication and Networking Technologies (ICCCNT), 2017 8th International Conference on*, pp. 1-6. IEEE, 2017.

[122] Noureen, Subrina Sultana, Vishwajit Roy, and Stephen B. Bayne, "Phasor measurement unit integration: A review on optimal PMU placement methods in power system." In *Humanitarian Technology Conference (R10-HTC), IEEE Region*, Dhaka, Bangladesh, *Vol. 10*, pp. 328-332, 2017.

[123] Aydemir, M. Timur, Ali Shan, and Abdul Karim Mesbah, "Optimum Placement of PMUs in the Power Transmission System of Afghanistan." *Gazi University Journal of Science,* Ankara Turkey, Vol. 30, No. 4 pp: 268-281, 2017.

[124] Ramírez-P, Sindy L., and Carlos A. Lozano, "Comparison of PMU Placement Methods in Power Systems for Voltage Stability Monitoring." *Ingeniería y Universidad*, Columbia, Vol. 20, No. 1, pp: 41-61, 2016.

[125] Raju, V. Vijaya Rama, and SV Jayarama Kumar, "An optimal PMU placement method for power system observability." In *Power and Energy Conference at Illinois (PECI), 2016 IEEE*, Telangana, India, pp. 1-5. IEEE, 2016.

[126] Singh, Satyendra Pratap, and S. P. Singh, "Optimal PMU placement in power system considering the measurement redundancy." *Int. J. of Advances in Electronic and Electric Engineering*, Varanasi, India, Vol. 4, No. 6, pp: 593-598, 2015.

[127] Paudel, Jyoti, Xufeng Xu, Karthikeyan Balasubramaniam, and Elham B. Makram, "A strategy for PMU placement considering the resiliency of measurement

REFERENCES

system." *Journal of Power and Energy Engineering,* Clemson, USA, Vol.3, No. 11, pp: 29, 2015.

[128] Manousakis, Nikolaos M., George N. Korres, and Pavlos S. Georgilakis, "Taxonomy of PMU placement methodologies." *IEEE Transactions on Power Systems, Vol.* 27, No. 2, pp: 1070-1077, 2012.

[129] Roy, BK Saha, A. K. Sinha, and A. K. Pradhan, "An optimal PMU placement technique for power system observability." *International Journal of Electrical Power & Energy Systems,* Kharagpur, India, *Vol.* 42, No. 1 pp: 71-77, 2017.

[130] Azizi, Sadegh, Ahmad Salehi-Dobakhshari, S. Arash Nezam-Sarmadi, and Ali Mohammad Ranjbar, "Optimal PMU Placement by an Equivalent Linear Formulation for Exhaustive Search." *IEEE Trans. Smart Grid* Vol. 3, No. 1, pp: 174-182, 2012.

[131] Ahmadi, A., Y. Alinejad-Beromi, and M. Moradi, "Optimal PMU placement for power system observability using binary particle swarm optimization and considering measurement redundancy." *Expert Systems with Applications*, Nowshahr Branch, Iran, Vol. 38, No. 6, pp: 7263-7269, 2011.